THE INTERNET OF THINGS

WIRELESS NETWORKS AND MOBILE COMMUNICATIONS

Dr. Yan Zhang, Series Editor
Simula Research Laboratory, Norway
E-mail: yanzhang@ieee.org

Unlicensed Mobile Access Technology: Protocols, Architectures, Security, Standards and Applications
Yan Zhang, Laurence T. Yang and Jianhua Ma
ISBN: 1-4200-5537-2

Wireless Quality-of-Service: Techniques, Standards and Applications
Maode Ma, Mieso K. Denko and Yan Zhang
ISBN: 1-4200-5130-X

Broadband Mobile Multimedia: Techniques and Applications
Yan Zhang, Shiwen Mao, Laurence T. Yang and Thomas M Chen
ISBN: 1-4200-5184-9

The Internet of Things: From RFID to the Next-Generation Pervasive Networked Systems
Lu Yan, Yan Zhang, Laurence T. Yang and Huansheng Ning
ISBN: 1-4200-5281-0

Millimeter Wave Technology in Wireless PAN, LAN, and MAN
Shao-Qiu Xiao, Ming-Tuo Zhou and Yan Zhang
ISBN: 0-8493-8227-0

Security in Wireless Mesh Networks
Yan Zhang, Jun Zheng and Honglin Hu
ISBN: 0-8493-8250-5

Resource, Mobility and Security Management in Wireless Networks and Mobile Communications
Yan Zhang, Honglin Hu, and Masayuki Fujise
ISBN: 0-8493-8036-7

Wireless Mesh Networking: Architectures, Protocols and Standards
Yan Zhang, Jijun Luo and Honglin Hu
ISBN: 0-8493-7399-9

Mobile WIMAX: Toward Broadband Wireless Metropolitan Area Networks
Yan Zhang and Hsiao-Hwa Chen
ISBN: 0-8493-2624-9

Distributed Antenna Systems: Open Architecture for Future Wireless Communications
Honglin Hu, Yan Zhang and Jijun Luo
ISBN: 1-4200-4288-2

AUERBACH PUBLICATIONS

www.auerbach-publications.com
To Order Call: 1-800-272-7737 • Fax: 1-800-374-3401
E-mail: orders@crcpress.com

THE INTERNET OF THINGS

From RFID to the Next-Generation Pervasive Networked Systems

Edited by

Lu Yan ✦ Yan Zhang

Laurence T. Yang ✦ Huansheng Ning

Auerbach Publications
Taylor & Francis Group
New York   London

CRC Press is an imprint of the
Taylor & Francis Group, an **informa** business

Auerbach Publications
Taylor & Francis Group
6000 Broken Sound Parkway NW, Suite 300
Boca Raton, FL 33487-2742

© 2008 by Taylor & Francis Group, LLC
Auerbach is an imprint of Taylor & Francis Group, an Informa business

International Standard Book Number-13: 978-1-4200-5281-7 (Hardcover)

Library of Congress Cataloging-in-Publication Data

The Internet of things : from RFID to the next-generation pervasive networked
    systems / Lu Yan ... [et al.].
        p. cm. -- (Wireless networks and mobile communications ; 8)
    Includes bibliographical references and index.
    ISBN 978-1-4200-5281-7 (alk. paper)
    1. Ubiquitous computing. 2. Radio frequency identification systems. 3.
Wireless communication systems. I. Yan, Lu. II. Title. III. Series.

    QA76.5915.I68 2008
    384.5--dc22   2007047411

Visit the Taylor & Francis Web site at
http://www.taylorandfrancis.com

and the Auerbach Web site at
http://www.auerbach-publications.com

Contents

Preface

With more than two billion terminals in commercial operation world-wide, wireless and mobile technologies have enabled a first wave of pervasive communication systems and applications. Still, this is only the beginning as wireless technologies such as RFID are currently contemplated with a deployment potential of tens of billions of tags and a virtually unlimited application potential. A recent ITU report depicts a scenario of "Internet of things" — a world in which billions of objects will report their location, identity, and history over wireless connections.

The realization of the "Internet of things" will probably require dramatic changes in systems, architectures and communications which should be flexible, adaptive, secure, and pervasive without being intrusive. Although the RFID technology has already laid a foundation for the "Internet of things," other research and development thrusts are also required to enable such a pervasive networking world, such as communications protocols, middleware, applications support, MAC, data processing, semantic computing and search capabilities, and even low-power technologies.

Significant R&D work has been undertaken over recent years on these systems, with pioneering work initiated in the US. In Europe, the European Union has been instrumental in supporting the many R&D facets of pervasive communications. Asia is also proactively moving into this field through various R&D initiatives on "ubiquitous communications." Emerging industrial interest in this field indicates that prospects for commercial applications of these technologies, are promising, and it is our hypothesis that a generic and comprehensive textbook is needed, where system-level problems in the context of the "Internet of things" are indicated and tutorials on their applications are required.

While placed in the specific context of the exciting expansion period of this research direction, this book will provide readers a comprehensive technical, practical, deploying, policy guidance covering fundamentals and recent advances in pervasive networked systems, from RFID towards "Internet of things."

The main features of this book include:

- The first book of its kinds to address the major new technological developments in the field "Internet of things"
- Reflects current research trends as well as industry needs
- A good balance between theoretical issues and practical issues
- Covers case studies, experience reports, and best practice
- Concept and technical issues addressed in this book are timely, and being seriously considered in the technology roadmap and strategies in EU, US, and Asia

This book serves well as a useful reference for students, educators, faculties, telecom service providers, research strategists, scientists, researchers, and engineers in the field of wireless networks and mobile communications.

We would like to acknowledge the effort and time invested by all contributors for their excellent work. All of them are extremely professional and cooperative. Our thanks also go to the anonymous chapter reviewers, who have provided invaluable comments and suggestions which help to significantly improve the whole text. Special thanks go to Richard O'Hanley, Jessica Vakili and Jay Margolis of Taylor & Francis Group for their support, patience and professionalism given in the whole publication process of this book. Last but not least, special thanks should also go to our families and friends for their constant encouragement, patience and understanding throughout this book project.

Lu Yan, Yan Zhang, Laurence T. Yang, Huansheng Ning

Contributors

Vishal Agarwal
Indian Institute of Technology
Electrical Engineering Department
Bombay, India

Paul Beardsley
Mitsubishi Electric Research Labs
Cambridge, MA, U.S.A.

Yolande Berbers
Katholieke Universiteit Leuven
Departement of Computer Science
Leuven, Belgium

Paul Brebner
National ICT Australia Limited
Braddon, Australia

J.T. Cain
University of Pittsburgh
Pittsburgh, PA, U.S.A.

Wenqing Cheng
Huazhong University of Science
    and Technology
Wuhan, China

William Cheung
University of Hong Kong
E-business Technology Institute
Hong Kong, China

Chi-Cheng Chu
University of California, Los Angeles
WINMEC
Los Angeles, CA, U.S.A.

Ruth Conroy-Dalton
University College London
The Bartlett
London, UK

Paul Dietz
Mitsubishi Electric Research Labs
Cambridge, MA, U.S.A.

Juan M. Estevez-Tapiador
Carlos III University of Madrid
Computer Science Department
Madrid, Spain

Eirini G. Fragkiadaki
Agricultural University of Athens
Faculty of Animal Science
    and Aquaculture
Athens, Greece

Rajit Gadh
University of California, Los Angeles
WINMEC
Los Angeles, CA, U.S.A.

Hector Gonzalez
University of Illinois at
 Urbana-Champaign
Department of Computer Science
Urbana, IL, U.S.A.

Stephen Hailes
University College London
Department of Computer Science
London, UK

Jiawei Han
University of Illinois at
 Urbana-Champaign
Department of Computer Science
Urbana, IL, U.S.A.

Peter J. Hawrylak
University of Pittsburgh
Pittsburgh, PA, U.S.A.

Jianhua He
Huazhong University of Science
 and Technology
Wuhan, China

Julio C. Hernandez-Castro
Carlos III University of Madrid
Computer Science Department
Madrid, Spain

Paris Kitsos
Hellenic Open University
School of Science and Technology
Patras, Greece

Giorgos Kostopoulos
University of Patras
Department of Electrical and
 Computer Engineering
Patras, Greece

Odysseas Koufopavlou
University of Patras
Department of Electrical and
 Computer Engineering
Patras, Greece

Jiming Liu
University of Hong Kong
E-business Technology Institute
Hong Kong, China

Wei Liu
Huazhong University of Science
 and Technology
Wuhan, China

Irene Lopez de Vallejo
University College London
The Bartlett
London, UK

Zongwei Luo
University of Hong Kong
E-business Technology Institute
Hong Kong, China

Zakaria Maamar
Zayed University
College of Information Technology
Dubai, U.A.E.

M.H. Mickle
University of Pittsburgh
Pittsburgh, PA, U.S.A.

Vasileios A. Ntafis
Agricultural University of Athens
Faculty of Animal Science
 and Aquaculture
Athens, Greece

Charalampos Z. Patrikakis
National Technical University
 of Athens
Department of Electrical Engineering
 and Computer Science
Athens, Greece

Zhao Peng
Huazhong University of Science
 and Technology
Wuhan, China

Alan Penn
University College London
The Bartlett
London, UK

Pedro Peris-Lopez
Carlos III University of Madrid
Computer Science Department
Madrid, Spain

B.S. Prabhu
University of California, Los Angeles
WINMEC
Los Angeles, CA, U.S.A.

Davy Preuveneers
Katholieke Universiteit Leuven
Departement of Computer Science
Heverlee, Belgium

Ramesh Raskar
Mitsubishi Electric Research Labs
Cambridge, MA, U.S.A.

Arturo Ribagorda
Carlos III University of Madrid
Computer Science Department
Madrid, Spain

Quan Z. Sheng
University of Adelaide
School of Computer Science
Adelaide, Australia

Nicolas Sklavos
University of Patras
Patras, Greece

Xiaoyong Su
University of California, Los Angeles
WINMEC
Los Angeles, CA, U.S.A.

C.J. Tan
University of Hong Kong
E-business Technology Institute
Hong Kong, China

Kerry L. Taylor
Commonwealth Scientific and
    Industrial Research Organisation
Information Engineering Laboratory
Canberra, Australia

Jeroen van Baar
Mitsubishi Electric Research Labs
Cambridge, MA, U.S.A.

Edward C. Wong
University of Hong Kong
E-business Technology Institute
Hong Kong, China

Eftychia M. Xylouri-Fragkiadaki
Agricultural University of Athens
Faculty of Animal Science
    and Aquaculture
Athens, Greece

Yan Zhang
Simula Research Laboratory
Lysaker, Norway

S.J. Zhou
University of Hong Kong
E-business Technology Institute
Hong Kong, China

Chapter 1

RFID Tags

Peter J. Hawrylak, M.H. Mickle, and J.T. Cain

Contents

Radio Frequency IDentification (RFID) has a long history and is part of the technological revolution both current and past. RFID enables quick payment of tolls and quick identification of items. In addition, RFID provides benefits, such as tracking assets, monitoring conditions for safety, and helping to prevent counterfeiting. RFID plays an integral part in the technological revolution along with the Internet and mobile devices, which are connecting the world together. This chapter focuses on the RFID tag and provides an overview and history of the various types of tags, their uses, and the physics behind their operation.

1.1 Introduction

RFID has a long history. RFID uses radio waves, which are one form of electromagnetic waves. As such, the genesis of RFID must be attributed to the founders of the electromagnetic wave theory: Michael Faraday, James Maxwell, and Heinrich Hertz. In the mid-nineteenth century, Faraday discovered that a current flowing through a wire created a magnetic field and conversely that, when a wire is exposed to a magnetic field, a current is present in the wire. Today this discovery is known as Faraday's law and along with the Ampere–Maxwell law forms the basis for a magnetic field, or near-field RFID systems. Maxwell developed the mathematical theory describing electromagnetism using the work of Faraday and others. Maxwell's work dealt only with the visible, infrared, and ultraviolet bands of the electromagnetic spectrum because the other types of electromagnetic waves were not known to exist [1]. Hertz was able to verify Maxwell's work and discovered radio waves [1], [2]. Today, the existence of numerous other electromagnetic (EM) waves, such as x-rays and gamma rays are known.

From these beginnings, scientists and inventors, such as Reqinald Fessenden and Guglielmo Marconi began in the early part of the twentieth century to develop many radio-based applications we use each day. By the 1930s, crystal radio sets became common place in the home allowing people to listen to music from far away cities and keep track of the local news and sporting events in real-time. Franklin D. Roosevelt was one of the first politicians to use the radio to further his political career, being elected President of the United States for four consecutive terms. The crystal radio with a headset instead of a free-standing speaker did not need electricity from the home to operate. The radio wave provided enough energy to move the diaphragm in the headset to reproduce the sound represented by the radio signal, although the listener needed to wear the headset to hear the radio broadcast. As electricity was not widely available outside of major cities in the 1930s, the crystal set was an important achievement. The ability of the crystal set to harvest energy from a radio signal is one of the key foundations of the passive RFID technology employed and being developed today.

The development of the radio also led to the development of Radio Detection And Ranging or radar. Radar utilizes the fact that radio waves reflect off an object enabling their range, height, and bearing to be determined. The militaries of the various world powers were all racing to develop radar technology in the years before and during World War II. Radar was employed to great effect by the British during the Battle of Britain. As World War II progressed, the British and the United States developed methods to reduce the size of the radar sets allowing them to be mounted in airplanes. Further, the minimum size of the target that could be detected by the radar set was reduced. Both advancements were key to finding and destroying German U-boats (submarines) to safeguard Allied shipping during the Battle of the Atlantic. Without radar and its advancements, the RFID systems using the far-field system would likely not exist today.

Another advancement in radio communications is the airplane transponder and its military counterpart, the Identify Friend or Foe (IFF) systems. These systems communicate with base stations, such as observation points or airplane control towers, to provide real-time monitoring and identification of airplanes. IFF systems are used by the military to distinguish between friendly and hostile forces to prevent friendly fire casualties and can be classified as active RFID systems. Today, active RFID systems are employed in similar applications and areas.

As with other electronic devices, advancements in integrated circuit (IC) fabrication were critical to the development of RFID. The first computers, made of vacuum tubes, encompassed entire rooms. Today with the advancement in IC fabrication, computers many times more powerful can be fit into a small device, such as a cell phone or a Personal Data Assistant (PDA).

The same advancements are true for RFID tags. Most RFID tags have an identifier, which must be stored in nonvolatile memory to retain the identity information of the tag when the tag is not powered. This is important in passive RFID tags as most of the time the tag is not powered. The digital component of a RFID tag made from vacuum tubes would be too large to be attached to anything but the largest assets, such as planes, trains, or cars. With the IC revolution, it was possible to shrink the size of the digital components. Current IC technology enables digital and analog components to be contained in the same physical chip (die). Today the digital components and the analog circuitry of a RFID chip are contained in a square chip no more than 1 mm on a side. The antenna is, by far, the largest part of a RFID tag today. Without the IC revolution, applying RFID tags to pallets, cases, and items would be unlikely if not impossible.

Equally important to the reduction of size is the reduction in the energy needed for an IC to operate. As passive RFID devices must harvest energy from the signal transmitted by the interrogator to operate, this reduction in size is critical. In the past 10 to 20 years, advancements in IC design and fabrication have allowed low power IC chips for RFID tags to be efficiently manufactured. Because the maximum theoretical power that can be harvested is inversely proportional to the distance between the interrogator and the tag, lowering the amount of energy required for IC operation increases range. Range is one critical metric for RFID systems especially in the supply chain where the goal is to read all tags on a pallet. Without sufficient range those tags in the center of the pallet may not be read.

Tracking animals was one of the earliest applications of RFID [2]. Probably the most wide spread use of a RFID system is the electronic article surveillance (EAS) tags. EAS tags are affixed to items in a store that cause an alarm to go off if customers enter or leave the store when the tags are in the active state. The EAS tags are deactivated during checkout and are designed to prevent theft of store merchandise. Checkpoint and Sensormatic, two leading manufacturers of EAS tags, were created in the 1960s [2].

Automatic toll payment is another early application of RFID. The key advancement in automatic toll payment was the common interface of the E-Z Pass system where users could use the same tag to pay tolls on transportation systems (roads, tunnels, and bridges) operated by different transit authorities. This compatibility was crucial, especially in the mid-Atlantic area of the eastern United States where a person traveling from Washington D.C. to New York City may need to pay three or more tolls with each toll going to different transit authorities. The compatibility of the E-Z Pass system can be compared to the compatibility of credit card payment machines that allow customers to use any major credit card. The E-Z pass system contributed two advancements to RFID: the need for compatible systems and the ability to read a moving E-Z Pass tag. These advancements are critical foundations for the electronic product code (EPC) RFID protocols. Reading of a moving tag is critical because most distribution centers that service retailers use conveyor belt systems to move and sort items. Reading these items as they move is much more efficient and profitable than periodically stopping the conveyors to read the RFID tags affixed to the items.

RFID is a method to transmit information without a direct hardware connection and without a line of sight between the two parties. Because of the lack of the need for a direct connection or a line of sight, RFID is often touted as a replacement for the barcode. The lack of a line of sight requirement means that a RFID system could take an inventory of an entire pallet simply by passing the pallet through a set of RFID interrogators. A barcode system could perform the same task provided that the barcodes on all products in the pallet are facing outward. This solution has significant limitations as typical item sizes are often too small to allow the barcodes on all items to face the outside of the pallet without leaving unused space in the middle of the pallet. RFID has another advantage over the barcode in that data can be written to RFID tags once deployed. Similar barcode technology exists in second dimensional (2D) barcodes, or data matrices, but this technology requires a barcode printer to print each additional piece of information. This is expensive, time consuming, and requires a very large area for the 2D barcode if a large amount of information is going to be stored in 2D mode. With a 2D barcode, the amount of data stored in the barcode is proportional to the area of the 2D barcode. RFID tags can contain a large amount of memory in a small physical area. The primary roadblock to storing huge amounts of data in a RFID tag is in harvesting enough energy to power the memory. Adding memory requires more energy for the tag to operate, reducing either read range or tag lifetime.

Most publications concerning RFID in the popular media are about RFID as a replacement of or an improvement over the barcode. This is the initial goal of many major retailers, such as Wal-Mart and Metro. While the RFID tag offers many improvements over the barcode, a number of major roadblocks still exist before RFID tags can be placed on the majority of individual items. Of primary concern is the cost of a single RFID tag and the ability to be 100-percent confident that the RFID tag will be read during any interrogation process. Current projects using RFID tags to track high-value items and assets have generated a return on investment (ROI), but the general case ROI is still not present [11].

Generating a return on investment (ROI) with an RFID system is critical to the widespread deployment of RFID. The system is used in numerous hospitals to identify patients for prevention of such things as giving one patient another patient's medications. It can also track supplies for inventory management and billing purposes. Further, many casinos use RFID systems to identify and track playing chips and cards [3]. In the hospital applications, RFID improves patient safety, generating a priceless ROI and, in the case of casinos, adds another deterrent for cheaters and forgers who forge playing chips hoping to exchange the fake chips for cash.

One area of major advantage for RFID is the ability to write information to RFID tags providing a key benefit in the fight against counterfeit drugs. There are numerous cases of patients unknowingly receiving and/or using counterfeit drugs, leading to potentially fatal results. The Federal Drug Association (FDA) is the regulatory body in the United States responsible for regulating the pharmaceutical industry. They have recently imposed a requirement for an e-Pedigree for all prescription medications, which is a record of every point of handling in the supply chain for that particular medication and used at the receiving end to verify the authenticity of the medication by verifying the chain of custody. The FDA mandate does not specify that RFID must be used, but only requires an electronic record of this information. RFID tags are one possible solution to the e-Pedigree mandate as they can be written to and updated during transit through the supply chain. Each RFID tag can contain the information about who transported the medication and when they took possession of the medication. This information, along with the unique identification assigned to each RFID tag, can be used to verify that the medication is not a counterfeit. In this application, RFID will help safeguard the public.

1.1.1 RFID Basics

All RFID systems contain three basic components. The first is the RFID tag that is attached to an asset or item. The tag contains information about that asset or item and also may incorporate sensors. The second component is the RFID interrogator, which communicates with (also called interrogating) the RFID tags. The third component is the backend system, which links the RFID interrogators to a centralized database. The centralized database contains additional information, such as price, for each RFID tagged item.

RFID tags generally fall into one of four categories: (1) passive, (2) active, (3) semiactive, and (4) semipassive. The semiactive and semipassive categories are currently somewhat grey areas in that they are very similar to either the active or passive categories, respectively, and often overlap. In this chapter, the following definitions will be applied to the above four categories of RFID tags. Passive tags are defined (in this chapter) as having no battery or onboard power source and communicates through backscatter. Active tags are defined as having an onboard power source, usually a battery, and having a powered receiver and transmitter. The powered receiver and transmitter allow for reception of very weak signals and transmission of signals over a long distance or through interference. Semiactive tags are those tags having an onboard power supply powering a microchip (intelligence), a transmitter, and a passive receiver. A semipassive tag is defined as having an onboard power supply powering only the microchip (intelligence), a passive receiver, and uses backscatter to communicate.

1.1.2 Passive RFID Tag Basics

Passive tags receive the most publicity and are currently being used by large retailers such as Wal-Mart and Metro to track inventory, and by the U.S. Department of Defense (DoD) to track supplies. Passive tags do not contain an onboard power source and derive all of the energy required for operation from the RFID interrogator interrogation signal. Thus, passive RFID tags have an unlimited lifetime with respect to power, but physical damage can still render passive RFID tags useless.

Passive tags contain a unique identification number, which is similar to a Universal Product Code (UPC) in concept, but provides additional information beyond a simple UPC. The UPC is a special barcode having a standardized appearance to allow any UPC scanner to be able to read the code. UPC found on many groceries and retail merchandise are read at checkout by the UPC scanner, which then matches a price to each code. UPC simply provides information about what the product is, for example, a can of tomato soup or a pair of blue jeans. The unique identifier contained in a passive RFID tag takes this one step farther and can identify which can of tomato soup, (e.g., tomato soup number 45362) or which pair of blue jeans (e.g., blue jeans number 86203) is being purchased. This is out of necessity because the current UPC-based systems work by scanning each item individually.

With RFID, a portal can be set up where a customer simply pushes a shopping cart through the portal and the value of the items in the cart is automatically totaled. Because each tag may be read multiple times during this process, it is necessary that each tag have a unique identifier. Otherwise, the customer would be charged multiple times for the same item. There are many benefits that accompany the unique identification of an item. For example, in the case of a recall of food items, health authorities can use this information to determine quickly where the bad food was sent and where it is being sold.

1.1.3   Active RFID Tag Basics

Active RFID tags also contain unique identifiers and may contain other devices, such as sensors. Active RFID tags comprise the second largest group (after passive tags) in use today. Active RFID tags have a powered or active transmitter and receiver that allow them to communicate over a greater distance and through more interference than passive tags. One major source of interference for passive RFID tags is metal and, in environments with high amounts of metal, such as shipping containers, active RFID tags are often used. Because active tags have an onboard power supply, they can incorporate sensors to monitor the environment. This is useful for monitoring food or drug shipments to verify that the contents were kept at the specified environmental conditions. Active tags produced by Savi (subsidiary of hackheed Martin) incorporate sensors to detect when a shipping container has been opened. TransCore (unit of Roper Industries) manufactures an active tag that is used to record the odometer readings of vehicles to allow trucking companies to easily track the odometer readings of their trucks to determine when scheduled maintenance is required.

1.1.4   Semipassive RFID Tag Basics

Semipassive tags use an onboard power supply to power the controller or microchip and may contain additional devices, such as sensors. Semipassive tags communicate using backscatter, and can communicate over a longer range than passive tags because they can use all the energy of the interrogator interrogation signal for communication. A passive tag must use part of the energy of the interrogator interrogation signal to power the controller or microchip. Semipassive RFID tags are useful in situations where reading is not difficult (i.e., no or limited metal) and where onboard sensors are needed to monitor an asset. Semipassive tags are still in the development phase and deployments are limited. One possible use of semipassive tags is to track pallets and to monitor the environment to which the pallets are exposed.

1.1.5   Semiactive RFID Tag Basics

Semiactive RFID tags can be thought of as an active RFID tag without an active (powered) receiver and as a result are often lumped into either active RFID tags or combined with semipassive RFID tags. In this chapter semiactive tags are discussed in the Current Outstanding Problems portion of the Active RFID tags section. Semi-active RFID tags have an active (powered) transmitter enabling their transmission to be detected at a greater distance or through more interference than a semipassive or passive RFID tag. Semiactive RFID tags are useful for tracking items in extremely noisy environments that prevent passive or semipassive tags from communicating with the reader. The nanoTag and burst switch, developed by the University of Pittsburgh RFID Center of Excellence and described later in this chapter (see section on active RFID tags), is one example of a semiactive RFID tag.

1.2   Passive Tags

1.2.1   How Backscatter Communication Works

All objects reflect radio waves, or radio frequency (RF) energy, and these reflected waves are the basis for pulsed radar systems. A pulsed radar system detects objects by first sending out a burst of RF energy and then waiting for the reflection to return. The difference in time between the radar station emitting the RF energy and the time the reflection was received, along with the speed of the RF wave in air is used to determine distance. Direction can be achieved by using a directional antenna as a part of the radar system. Two basic types of antenna exist, omnidirectional and directional. An omnidirectional antenna emits RF energy in all directions, while a directional antenna emits a narrow band of RF energy in a specific direction.

Because all objects, including RFID tags, reflect RF energy, backscatter takes advantage of this. A passive RFID tag contains an antenna that is used for two purposes: (1) the antenna is used to harvest energy from the interrogator continuous wave (CW) RF signal, and (2) the antenna is used by the tag to communicate to the interrogator by modifying reflections. The amount of energy that the tag can harvest depends on many factors, but the distance between the interrogator and tag, the interrogator transmitter power, and the efficiency of the tag antenna are the major factors. The efficiency of the RFID tag antenna is determined by the quality of the matching between the antenna and the tag circuitry.

The quality of matching between the antenna and the tag circuitry determines how effectively the energy can be transferred between the antenna and tag circuitry. A matched system is defined as a system with an antenna having the same resistance and opposite reactance as the tag output circuitry. Impedance is a complex value where the real part is the resistance and the imaginary part is either capacitive or inductive. In an RFID tag the maximum power transfer is achieved when the impedance is purely resistive and the value of the antenna resistance matches the resistance of the tag chip. The second part, matching the resistance of the antenna to the resistance of the tag chip, or load, is critical to achieve maximum power transfer between the antenna and the tag chip. Thus, the tag circuitry must have an impedance that cancels out the imaginary part of the antenna impedance to achieve maximum energy transfer. Matching the antenna and the chip on a RFID tag is a complex process. Further, the fact that the matching is affected by the item to which the tag is attached makes matching very difficult for general RFID use. For this reason, a number of different antennae are designed using the same RFID chip, with each antenna design applied to a specific type of item.

When a passive RFID tag is powered by an interrogator interrogation signal, the tag can alter the matching between the antenna and the tag chip. This is normally achieved by a switch that is controlled by the tag circuitry to switch in an additional capacitance in parallel with the antenna to detune the antenna. When the antenna and tag chip are matched, with the additional capacitor not in the circuit, the RFID tag reflects an amount of RF energy, A. With the capacitor in the circuit the RFID tag reflects an amount of RF energy, B. The amount of energy reflected, A, and B, are different. By switching the capacitor in and out of the circuit, the RFID tag can modulate the RF energy it reflects back to the interrogator. A predefined backscatter protocol allows communication from the tag to the interrogator.

1.2.2 Operating Frequencies: An Overview

Passive RFID tags have been in use for some time. They operate in three frequency bands: (1) low frequency (LF), (2) high frequency (HF), and (3) ultra high frequency (UHF).

Low frequency RFID tags primarily operate at 125 kHz, but the operating frequency can range from 30 kHz to 300 kHz [4]. Low frequency RFID tags are commonly used to track pets and they work rather consistently when applied to metal or in free air. These tags are normally encased in a glass cylinder and then injected under the pet's skin. However, the range of an LF RFID tag is relatively short and the maximum data transfer rate is slow.

Recall that backscatter-based communication utilizes the RF wave reflected from the tag to communicate with the interrogator. The interrogator emits a sine wave of energy where the tag modulates the reflection. The frequency of the sine wave emitted by the interrogator is the operating frequency of the system and this relates to the maximum data rate of the system. Hence, for an interrogator and tag operating at 915 MHz (UHF), the tag could modulate theoretically a maximum of millions of bits of data per second. A system operating at 125 kHz (LF) can modulate thousands of bits of data per second. These numbers assume that the system can modulate and distinguish virtually every period in the sine wave, but, in both cases, multiple periods are modulated together to form a single bit. Therefore, the data rates of UHF, HF, and LF systems are much lower than the maximums described above.

High frequency RFID tags can operate between 3 and 30 MHz, but operate primarily at 13.56 MHz [4]. High frequency RFID systems are employed in numerous libraries, including the Vatican in Rome. The Federal Communications Commission (FCC), which governs nongovernment use of the radio spectrum in the United States, has defined a range of frequencies 13.56 MHz +/- 17 kHz as one of the bands for Industrial Scientific and Medical (ISM) use [5]. This band can be used in the United States and worldwide without a license and is the main reason for HF RFID to be centered around 13.56 MHz. License-free frequency bands are important to reduce costs as most frequency bands are set aside for the public good (e.g., maritime radio frequencies) or are licensed, for a substantial fee, to private groups (e.g., radio stations).

There are a number of different International Standards Organization (ISO) and proprietary standards for HF RFID systems operating at 13.56 MHz. Thus, the frequency (13.56 MHz) is available worldwide, but the systems can speak one (or possible more) of several different languages (different ISO standards). Hence, HF RFID can be used only if all entities wishing to use the system agree on the communication standard to use or if the various standards used are compatible. High frequency RFID can be read at a distance greater than LF RFID, but at a distance shorter than UHF RFID. Similarly the data rate of HF RFID is greater than LF RFID, but less than UHF RFID. However, HF RFID has better resistance to interference from things, such as metal, than UHF RFID, but is more susceptible to interference than LF RFID.

Ultra high frequency RFID operates primarily at frequencies between 866 MHz and 960 MHz, but UHF RFID could be considered anything between 300 MHz and 3 GHz [4]. The Gen-2 protocol is the primary protocol used by UHF RFID tags and operates between 866 MHz and 960 MHz. The Gen-2 protocol is accepted in most of the world, but no common frequency is available worldwide for UHF RFID tags. For this reason, most of the publicity in the retail sector concerning RFID usually references Gen-2 and UHF RFID systems. Thus, UHF RFID systems are distinguished from HF RFID systems with respect to operating frequencies and protocol.

Ultra high frequency RFID systems have a common language (protocol), Gen-2, but do not have a common operating frequency. For example, in the United States, UHF RFID systems operate between 902 MHz and 928 MHz, while in Europe, UHF RFID systems are allocated the frequency bands 860 and 870 MHz. In Japan, Gen-2 operates at approximately 960 MHz.

Due to the higher frequency of UHF RFID systems, the data rates of these systems are much higher than LF and HF RFID systems. The read range of UHF

RFID systems is also greater than both LF and HF RFID systems, but UHF RFID systems are greatly affected by substrate interference, especially metal. Due to these systems susceptibility to interference of the substrate, special care must be taken when applying UHF RFID tags to items and they may not function on some items.

European regulators require that RFID systems listen before transmitting (listen before talk) in an attempt to reduce collisions and general interference with other wireless systems. This is similar to the requirement of 802.11 (WiFi)-enabled devices to listen for a specified time without hearing any other transmission before sending their transmissions. The "listen before talk" requirement reduces the maximum available data rate of European UHF RFID systems, but the data rate of these systems is still faster than LF and HF RFID systems.

Read rate is very important in RFID systems used to track inventory because it determines the maximum speed of a conveyor belt in a warehouse or distribution center. The faster the conveyor belt moves, the quicker product can be processed and loaded onto trucks, reducing the cost of delivery. A similar argument is also valid for production facilities that employ RFID, as orders must be filled within the specified timeframe. Passive RFID tags communicate using one of two methods (SAW-based tags and chipless RFID tags are not covered in this chapter): near-field and far-field. The near-field and the far-field use a different mechanism for communication and to power the RFID tag.

1.2.3 Magnetic Coupling: Near-Field

The near-field RFID tags utilize magnetic coupling, which is the same principle as is used in a transformer. In an electrical transformer, there are two coils, the primary and the secondary. Current passes through the primary coil and, as a result of Faraday's law of induction, induces a current in the secondary coil. Transformers are used by the electric utilities to reduce the voltage from the high voltage used in transmission lines to a safer low voltage for use in the home. Transformers are used in many other applications, but the above is a very common use.

In a near-field RFID system, the interrogator has an antenna that acts as the primary coil in the transformer described above. The near-field RFID tag acts as the secondary coil in the transformer described above. Near-field RFID tags derive power from the induced current due to the magnetic field generated by the interrogator, the primary coil.

The amount of energy delivered to the RFID tag is dependent on the strength of the magnetic field and the number of lines of flux that pass through the RFID tag antenna. The magnetic field generated by the interrogator (primary coil) consists of lines of flux forming closed circles centered around the interrogator antenna (primary coil). Hence, a cylindrical-shaped magnetic field exists, encasing the interrogator antenna consisting of a large number of lines of flux. The more of these lines of flux that pass through the RFID tag antenna, the greater the current that is induced in the RFID tag, resulting in a greater amount of energy for the RFID tag.

1.2.4 Electromagnetic Coupling: Far-Field

Far-field RFID tags utilize electromagnetic coupling in what is termed the far-field. There are several different definitions of the beginning of the far-field, but the one commonly used is

$$r = \frac{\lambda}{2\pi} \qquad (1.1)$$

where *r* is the distance (in meters) from the emitting antenna where the far-field begins and λ is the wavelength (in meters per second) of the signal [4].

1.2.5 Near-Field and Far-Field: Some Key Points

The near-field extends essentially out to the beginning of the far-field. The effective distances for a near-field communication of a typical LF, HF, and UHF RFID systems calculated from Equation 1.1 are shown in Table 1.1. While the near-field is present in all three RFID systems, the maximum range of the near-field in an UHF RFID system is about 50 cm, while LF and HF RFID systems have a near-field of much greater range.

While the distance between the interrogator and the tag, or read distance, is important in an RFID system, of almost equal importance in many applications of RFID is the sensitivity to tag orientation. In applications where RFID is used on a conveyor belt system, as in a warehouse or distribution center, the location of the tag is known to be on one of the six faces of the box. This allows interrogators to be set up in a portal fashion to cover the four principle locations: top, bottom, left, and right. Modern warehouse systems provide the ability to locate the tag on one of those four mentioned sides. Hence, in a warehouse or distribution center, RFID tag orientation is not the major concern because the inherent infrastructure is able to orientate the RFID tag into one of a few set orientations.

In many other applications, the sensitivity of the system to RFID tag orientation is critical. One example of such a system is a RFID-enabled checkout line where the shopper saves time by just pushing the shopping cart through the RFID portal. In this example, the system works at nearly all the RFID tag orientations so as to be cost effective to the merchant (all items are accounted for) and to be time-efficient for the customer.

Table 1.1 Maximum Range of the Near-Field of LF, HF, and UHF RFID Systems

Frequency	Near-Field Range
125 kHz (LF)	381.97 m
13.56 MHz (HF)	3.52 m
915 MHz (UHF)	0.05 m

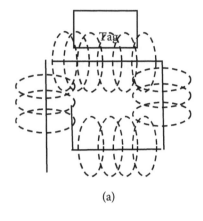

(a)

Tag 90 degrees to (a)

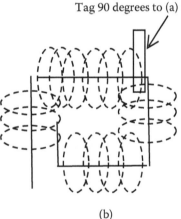

(b)

Figure 1.1 Lines of flux passing through RFID tags at two different orientations (a and b) 90 degrees apart.

Near-field and far-field RFID tags have different orientation sensitivities. Near-field RFID systems are less sensitive to the orientation of the RFID tag than are far-field RFID systems. This is because of the different methods by which RFID interrogators in each type of system deliver energy to the RFID tag.

Recall that the near-field RFID systems employ magnetic coupling to transfer energy from the interrogator to the tag by means of a cylindrical magnetic field generated around the interrogator antenna and that the amount of energy transferred to the tag depends on the number of lines of flux passing through the tag antenna. Imagine a square, spiral antenna attached to the interrogator and an RFID tag with a square, spiral antenna. If the tag antenna is placed in a plane parallel to the interrogator antenna, as shown in Figure 1.1a, several lines of flux pass through the tag antenna. Alternatively, assume that the tag antenna shown in Figure 1.1a is rotated 90 degrees to an orientation shown in Figure 1.1b. In this case, several lines of flux still pass through the tag antenna. Thus, in both cases, the tag antenna is magnetically coupled to the interrogator antenna allowing the tag to harvest energy from and to communicate with the interrogator. Hence, near-field RFID systems are not sensitive to tag orientation provided that the RFID tag antenna is within the magnetic field generated by the RFID interrogator antenna.

Conversely, far-field RFID systems are more sensitive to tag orientation. Recall that far-field RFID tags communicate by reflecting RF waves back to the interrogator. RFID tags derive their power by absorbing RF waves emitted by the interrogator. The RF waves emitted by the interrogator antenna break into several independent wave fronts. These wave fronts can interfere with each other. For instance, two wave fronts may cancel each other out, creating what is called a null, or they can constructively add together, effectively doubling the incremental electric field in that region. The commonly used dipole antenna generates a series of egg-shaped regions in which a RFID tag can be powered and read. The region where a RFID tag can be powered and read is generated by a software tool developed at the University of Pittsburgh RFID Center of Excellence and is illustrated in Figure 1.2 for a dipole antenna.

There are multiple of these egg-shaped regions, each with a center line at different angles from the normal vector to the dipole antenna. At some distance away from the interrogator, these egg-shaped regions allow effective reading of a RFID tag in only a limited portion of the total space. The illustration of these areas, shown in Figure 1.3, illustrate this point, as a RFID tag can only be read along the wall if it falls into one of the egg-shaped areas. Hence, the long range of far-field RFID systems may be limited to finding the multiple "sweet spots" where the tag is in the egg-shaped readability area.

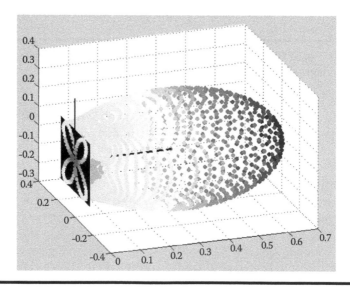

Figure 1.2   Radiation pattern.

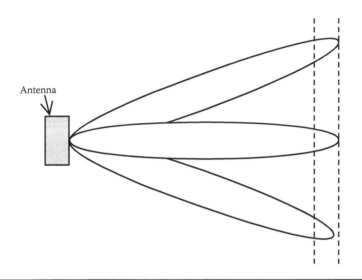

Figure 1.3 Simplified RFID reader radiation pattern.

Within the egg-shaped areas of readability, the orientation of the RFID tag in a far-field RFID system is critical. Because far-field RFID tags reflect the RF waves emitted by the interrogator back to the interrogator to communicate with the interrogator, the far-field RFID tag can be likened to a mirror and the RF wave likened to a beam of light. Imagine the tag as a square mirror with reflective material on all six faces (top, bottom, left, right, front, and back) with a length and height that is much greater than the width. If the mirror is placed perpendicular to the beam of light (RF wave) emitted by the interrogator, as shown in Figure 1.4a, a large amount of the light is reflected back to the mirror. Now, image spinning the mirror 90 degrees to the orientation, shown in Figure 1.4b. In this orientation, the mirror reflects only a small fraction of the light. The principle is the same as the reflection of RF waves, the greater the area that is shown to the RF wave, the greater the reflection determined by the radar cross section. With only a small area exposed to the RF wave, the tag has a small radar cross section and is either not able to harvest enough energy to power itself or to reflect enough of the RF waves to generate a strong enough signal to be detected by the interrogator.

Thus, the near-field RFID systems are more robust with respect to the tag orientation, but have a shorter read range than a far-field RFID system.

1.2.6 Manufacturing Issues with Passive RFID Tags

Passive tags are made from three major components: (1) the chip, (2) the strap, and (3) the antenna. The chip contains the digital and analog circuitry, which

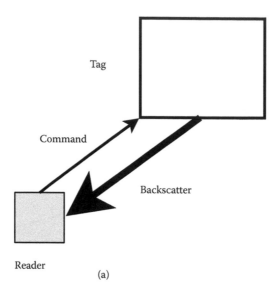

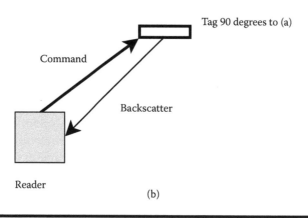

Figure 1.4   Backscatter communication and reflective mirror analogy.

control the chip and stores information, such as the tag ID. The chip contains very small pads (on the order of tens of microns on a side), which must be connected to the antenna to receive and backscatter information. The strap contains two large pads to which the small pads on the chip are attached. The strap provides a large bonding area to the antenna, which allow for faster assembly of the RFID tag and provide a better physical connection with the antenna, although an additional step is introduced into the manufacturing process.

The quality of the attachment of the chip to the strap and of the strap to the antenna is critical in the performance of the RFID tag. Variations, notably the placement, in either attachment process cause variations in the chip performance, most notably in the read-range of the chip. In extreme cases if the quality of one of the attachments is too poor, the RFID tag may not function at all.

The minimum roll diameter is another critical characteristic of a RFID tag. The minimum roll diameter is the minimum diameter of the cylinder that a RFID tag may be wrapped around and still function. The attachment process of the chip to the strap and the strap to the antenna is of critical importance here because it is often the case that these connections break when the tag is wrapped around a small cylinder. Reducing the minimum roll diameter requires strengthening or making these connections more resistant to bending. This is important in the pharmaceutical industry as most pill containers are cylindrical in shape and typically smaller than the minimum roll diameter.

1.2.7 The EPC Gen-2 Protocol

The Gen-2 protocol is often mentioned in conjunction with RFID. The Gen-2 protocol was developed by EPCglobal. One of the goals of EPCglobal is to provide a standardized method of communication between the interrogator and the tag to allow interoperability between interrogators and tags developed by different companies, and to provide a standard communication protocol for the entire world. The Gen-2 protocol is the second generation protocol and provides improvements over Gen-1. Currently most users of RFID have phased out Gen-1 tags and are currently using Gen-2 tags. The Gen-2 protocol was recently approved by the International Standards Organization (ISO) for an ISO standard; ISO 18000-6C.

The EPC Gen-2 protocol is based on a master–slave-style architecture, with the interrogators acting as the masters and the tags acting as the slaves. The interrogators initiate all communication and actually provide the power for the tags to operate. The Gen-2 protocol is half-duplex. Three major steps are required for an interrogator to access a tag. First, the interrogator must issue a Select command to pick the tag population to which it wishes to communicate. This feature of the Select command allows the interrogator to divide up a group of tags into smaller subgroups to reduce collisions, and reducing the time required to access one or more tags. Second, the interrogator must take an inventory to determine the EPC numbers of all tags in a given population. The Query (issued only once at the beginning of the inventory) and QueryRep commands are the main commands used to obtain this information. When receiving a Query command, each tag selects a random number for the initial slot value and beings reducing it according to the standard specifications. If the slot value is 0, the tag will respond transmitting a 16-bit random number uniquely identifying a tag within statistical parameters specified by the protocol. The interrogator then issues an ACK (acknowledgment) with the tag's 16-bit random number, causing the tag to transmit its protocol control,

EPC number, and cyclic redunancy check (CRC) values to the interrogator. The CRC value is used to detect errors, possibly from interference, present in a received message. The QueryRep command is a repeated query causing all tags that have not yet responded with their 16-bit random number to decrement their slot value by 1. If the slot value is 0, the tag will reply with its 16-bit random number and the interrogator will ACK that tag and obtain the protocol control, EPC number, and CRC values. Third, after obtaining the protocol control, EPC number, and CRC values the interrogator can communicate directly with the tag to perform more complex operations, such as reading or writing memory, locking or unlocking the tag, or killing the tag.

The physical link from the interrogator to the tag can function at different bit rates. Bit rates from 26.7 kbps up to 640 kbps are possible. This allows interrogators to communicate quickly with a tag under more favorable conditions, or slower to accommodate slower tags or in poor conditions. The variable data rate allowed in Gen-2 is an improvement over many Gen-1 systems, which supported only a single data rate preventing Gen-1 systems from adapting to the current environment [6]. Variable data rates are useful in allowing for the worldwide operation of the Gen-2 protocol. For example, the regulations governing RF transmission require that interrogators listen to ensure the channel is clear before transmitting, reducing the maximum data rate between the interrogator and tag. The adjustable data rate of Gen-2 enables it to be used worldwide much easier than the single data rate Gen-1 systems [6].

The signal is modulated using amplitude shift keying (ASK) in which the full signal amplitude indicates one symbol and the attenuated amplitude represents another symbol. The data is encoded using pulse interval encoding where the data is represented by the lengths of distinct pulses.

The tag to interrogator communication link also has variability. The tag must be capable of both ASK and phase shift keying (PSK). The interrogator selects the modulation, encoding, and bit rates. Two data encodings, FM0 and Miller, are possible. Bit rates can range from a low of 5 to 320 kbps for Miller encoding, and from a low of 40 to 640 kbps for FM0 encoding. The two encodings and variable bit rates allow the interrogator to optimize the communication to better suit the environment. Use of a Miller encoding results in a slower maximum data rate than an FM0 encoding, but the Miller encoding is more resistant to interference from RF noise than is an FM0 encoding [6]. In poor environments, slower bit rates may be the better option, but in ideal environments the faster bit rates will increase the number of tags that can be read in one second.

Gen-2 uses sessions to enable multiple interrogators to communicate with the same tag simultaneously. Up to four sessions are possible in Gen-2, hence, a single tag can communicate with four different interrogators simultaneously [6]. Without sessions the tag state could be updated by independent interrogators preventing all interrogators from communicating with the tag. Sessions are important because in most applications tags will be within range of multiple interrogators.

1.2.8 Current Outstanding Issues with Passive RFID Tags

Research and development is required in several directions to advance passive RFID tags. The five main areas today are: (1) the size of the tag, (2) the cost of the tag, (3) the read range, (4) the read rate, and (5) tag security.

1.2.8.1 Reducing Tag Size

The size of the RFID tag is important for item-level tagging because a small tag requires less area on the product. However, even though a tag may be made smaller, the minimum roll diameter is most critical in the area of the chip/antenna connection, which is unaffected by the reduced antenna size. This is important for smaller items or for items where the available area for a RFID tag is limited. The primary issue with reducing the size of the RFID tag is in reducing the size of the antenna. The ideal antenna length is on the order of the wavelength of the frequency at which the antenna is designed to operate. When reducing the size of the antenna classical dipole to get the best performance, the half-wavelength of the antenna must be fit to the tag area. While the antenna one-half wavelength in length could be fit to a given area, the geometry of the antenna layout may introduce additional capacitance or inductance throwing off the matching between the antenna and RFID tag chip. To achieve the best performance, the RFID tag chip must be rematched to the new antenna. This is a time-consuming and labor-intensive process, although it is done only once for each tag design.

1.2.8.2 Lowering Tag Cost

The second major research area is reducing the cost of the tag. Cost is extremely important in the retail setting at the item level. Current barcode and UPC technology has almost no additional cost because the barcode or UPC is simply printed somewhere on the label. RFID tags, on the other hand, require components that cannot currently be printed as part of the item label. Thus, RFID tags add an additional cost to the product, which must either be absorbed by the producer or passed on to the customer.

1.2.8.3 Increasing Read Range

The third major research area is in increasing the read range of the RFID tag. Here, read range is defined as the maximum distance between the interrogator and the tag such that the tag can still be read. Read range is determined primarily by three factors: (1) the amount of energy the tag can harvest from the interrogator CW signal, (2) the amount of energy required to operate the digital control logic, and (3) the ability of the interrogator to detect a response. The third factor is also dependent on the quality of the receiver used by the interrogator. As receiver technology continues to improve, the interrogator will be able to detect tag responses having a much weaker signal than is possible today. This will enable the interrogator to receive replies from tags that are farther away because the received power is inversely proportional to the distance raised to the same power (often "distance squared" is used).

1.2.8.4   Increasing Read Rate

The fourth major area of research is increasing the read rate available in an RFID system. Here read rate is defined as the number of RFID tag replies that can be successfully received and decoded by the interrogator in one second. Read rate is important because one of the main applications of passive RFID tags is to be used in place of or to complement barcodes. Currently conveyor belt systems are used at manufacturing facilities and warehouses. As the item moves through the conveyor belt system, the RFID tag attached to the item must be read before it leaves the interrogators area of readability, which is defined as the area within which the interrogator can successfully read a RFID tag 100 percent of the time. Thus, a faster read rate allows the conveyor belt system to operate at a faster speed, enabling the manufacture or processing of more products. Read rate also is important in areas of high tag density where collisions must be arbitrated.

1.2.8.5   Improving Tag Security

The fifth major area of research is in the security of RFID tags. They contain a unique identifier that identifies the tagged item. Because RFID tags can be read at a distance, even through walls, it is possible for a malicious individual, while remaining concealed, to read RFID tags on items owned by a person using a RFID interrogator. With the tag ID numbers, the individual can determine what items the other person is carrying. This could lead to a release of private information, such as the types of medication a person takes. More sinisterly, the individual could use the tag ID numbers to determine which person is carrying a valuable item, such as a diamond ring, and which person is carrying a low-priced item, such as a gallon of milk. The individual could then target the person with the diamond ring. Tracking is also possible because of the uniqueness of the tag ID numbers. The Gen-2 protocol defines a kill command that can be used to permanently disable the RFID tag. The drawback of killing the tag is that the information it contains is lost and cannot be accessed or used at a later time.

Another security problem is a thief using an RFID interrogator to either rewrite the tag ID number or, in the case of a Gen-2 tag, to simply kill the tag. Changing the tag ID number effectively changes the price of an item. In this case, the thief writes a tag ID number of a low-priced item to a high-priced item. Alternatively, a thief could simply issue the kill command to a Gen-2 RFID tag causing it not to reply to any interrogator. The kill command requires a tag password. There are many well-known attacks that could be used to determine the password. The key here is to make the time required to determine the password long enough to prevent a malicious individual from profiting from the password.

1.3 Active Tags

Active RFID tags are tags with an onboard power supply, typically a battery, with an active (battery-powered) transmitter and receiver, termed an active transmitter and active receiver, respectively. Active tags have a lifetime that is limited by the lifetime of the onboard power source. Typical lifetimes of commercially available active tags from Savi and TransCore are usually several years.

1.3.1 Active Communication Versus Backscatter Communication

An active receiver is capable of detecting and decoding very weak signals. Thus, active tags can successfully receive transmissions from interrogators that are much farther away or through more interference than can a passive RFID tag. Similarly, active transmitters have a much greater range than backscatter communication. Active transmitters typically have a range of between 100 and 500 m. Further, because active transmitters can emit a much stronger signal than backscatter tags, an active RFID tag can communicate with an interrogator over a longer distance or through more interference than a passive RFID tag. The RF reflections used in communication based on backscatter can transmit with a signal strength no greater than the strength of the signal received at the passive tag. Conversely, an active transmitter on an active tag can transmit at a specific power level regardless of the signal strength of the received signal (provided the onboard power supply has enough energy remaining).

1.3.2 Active Tags Conforming ISO 18000-7

The ISO 18000-7 standard specifies the communications link between an ISO 18000-7 compliant interrogator and an ISO 18000-7 compliant active tag. This section gives a brief overview of the available functionality of ISO 18000-7.

ISO 18000-7 systems communicate in a frequency band of 433.915 to 433.925 MHz and use frequency shift keying (FSK) modulation. Recall that passive Gen-2 tags use either ASK or PSK modulation through backscatter communication. ASK and PSK simply alter either the amplitude or phase (respectively) of the carrier wave (sine wave). Thus, it is easy for ambient RF noise or interference to introduce errors into the signal.

FSK uses multiple frequencies to encode different data values. ISO 18000-7 employs binary FSK, or 2-FSK, and uses only two frequencies to represent the Low and High symbols. In ISO 18000-7, the Low symbol is represented by the frequency 433.925 MHz and the High symbol is represented by the frequency 433.915 MHz. The receiver looks for these two frequencies. For noise to affect the system, there would have to be noise on both frequencies (433.925 and 433.915 MHz) and this is less likely than noise on a single frequency such as is used by ASK and PSK.

An ISO 18000-7 system is a Master–Slave-style architecture with the interrogators acting as the masters and the tags acting as the slaves. The interrogators initiate all communication and the tags respond only to interrogator inquiry and do not respond to overheard tag replies. Tags can be put to sleep to conserve energy and extend lifetime, and are wakened by a tone consisting of a 30 kHz FSK modulated pulse (centered at 433.92 MHz).

Communication between interrogator/tags and tag/interrogator consists of four parts. The first part is the preamble, which allows the receiver to lock onto the signal. The second part is the Sync pulse and it immediately follows the preamble. The Sync pulse informs the receiver if the message is from a tag or an interrogator. The third part immediately follows the Sync pulse and contains the data. The data is encoded using Manchester encoding, which encodes the clocking signal into the data signal. This allows for the receiver to reconstruct portions of the messages even when some of the portions arrive outside of the normal time window. In Manchester encoding, there is a transition in the middle of each bit time. In ISO 18000-7, a symbol High to symbol Low transition in the middle of the bit time denotes a logic 0, and a symbol Low to symbol High transition in the middle of the bit time denotes a logic 1. A stop bit is included at the end of every byte (8 bits equal 1 byte) to further improve error detection. The fourth part of the communication consists of the end pulse, which is a 36 μsec period of continuous symbol Low denoting the end of the message. An ISO 18000-7 communication originating from an interrogator is shown in Figure 1.5 illustrates a communication originating from a tag; note the difference in the lengths of the Sync pulses.

There are a number of commands specified by ISO 18000-7 falling into one of three categories. The first category consists of inventory commands, which instruct all tags to transmit their IDs to the interrogator. The commands in this first category serve the same purpose as the Query commands in the Gen-2 protocol. The second category consists of control commands. These commands are used to place the tag into the sleep state, the unsecured state, and the secured state. The third set of commands are the data commands allowing the tag password to be written, various parameters updated, and memory to be written and read.

As part of all tag replies, the tag includes a flag indicating if the battery is low. The interrogators can use this flag to keep a tag with a low battery asleep for longer

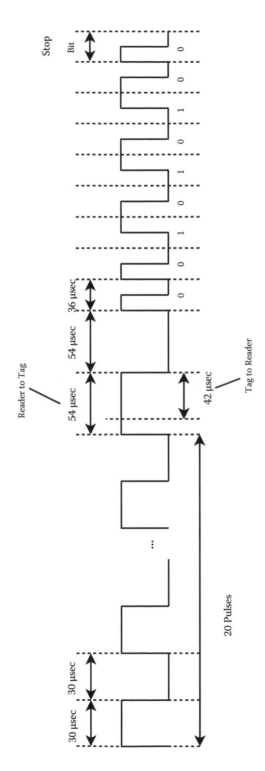

Figure 1.5 ISO 18000-7 communication signal; Reader to Tag communication uses the 54 μs sync pulse; and Tag to Reader communication uses the 42 μs sync pulse.

periods in an attempt to extend lifetime. Also included in all commands and replies is the Interrogator ID. The Interrogator ID can be used to identify each interrogator and can be used as a filter by interrogators when they receive replies. Interrogators receiving a reply with a different Interrogator ID than their own Interrogator ID can simply ignore that reply.

Another and possibly better use of the Interrogator ID is to use it as an Internet Protocol (IP) address in a network. In this case, an interrogator that receives a reply addressed to another interrogator could transmit that reply to the interrogator matching the Interrogator ID in the reply. If the interrogator the reply is addressed to did not receive the reply for some reason, by receiving the forwarded reply from the other interrogator, it saves the interrogator from retransmitting the same command. By retransmitting the command, the interrogator forces the tag to retransmit the reply; clearly the less the tag transmits the longer the battery will last. The system of forwarding overheard replies can conserve tag battery life and, therefore, extend the tag lifetime. One problem with this system is that if both interrogators received the reply, then the forwarded reply is duplicated and takes up bandwidth on the network connecting the interrogators.

1.3.3 Sensors

Because of the onboard power supply, active tags can perform tasks even when an interrogator is not interrogating the active tag. This is particularly useful if sensors are added to an active RFID tag. As previously mentioned in the introduction, active tags with accompanying sensors can be used to monitor conditions during shipment (i.e., temperature), can provide security to shipping containers (i.e., send alerts when the shipping container is opened), and can streamline and schedule periodic maintenance (i.e., monitor mileage on vehicles). The addition of sensors to active RFID tags opens the door to many applications that require both identification and monitoring of some condition. Many applications exist in the medical fields where identification of a patient is required while monitoring one or more conditions, which may be helpful in providing better care.

There are many producers of active RFID tags and many producers of sensors. With so many different families and types of active RFID tags and sensors, it is critical that a standardized interface exists between the active RFID tag and the sensor. For example, some sensors have a voltage output that is proportional to a specific phenomenon while others may provide a proportional current to that phenomenon. A standardized interface provides two key benefits: it reduces design time and it reduces cost. The International Electrical and Electronic Engineers (IEEE) standard IEEE 1451 family of standards provides a standardized sensor interface.

These IEEE standards were originally designed to connect a controller to multiple sensing devices. IEEE 1451.1 defined the standardized interface for the controller, termed a Network Capable Application Processor (NCAP) in the IEEE 1451 family of standards. The concept of a Transducer Electronic Data Sheet (TEDS) data sheet containing all information required for the NCAP to understand the sensor output reading and to convert that output reading into meaningful units. Transducer calibration information is also provided by the TEDS. The TEDS data sheet enables any NCAP to understand and convert the transducer (sensor) output into meaningful units. The interoperability provided by the NCAP and TEDS specification enables any controller with (or complying to) an NCAP interface to connect to any transducer implementing the TEDS and appropriate interface to the NCAP. This allows a sensor module developed by one manufacturer to be integrated in an easy and straightforward fashion into an active RFID tag produced by another manufacturer. Use of the IEEE 1451 family of standards also allows the reuse methodologies utilized in the system on a chip (SoC) design process to be applied to active RFID tags with built-in sensors.

1.3.4 Security

Security is an important issue in RFID systems. As previously mentioned, thieves can read RFID tags to target persons carrying high valued items, and clandestine readings of RFID tags on a person can allow that person to be tracked or to gain private information about him or her (i.e., what medications he or she takes) obtained through the RFID tags.

Encryption is commonly employed on the Internet to protect personal information, often in the form of credit card numbers. Encryption schemes, such as RSA (Rivest, Shamir, Adleman) and AES (advanced encryption standard) are very complex requiring both time and additional computing resources (software or hardware). Any encryption performed in a passive RFID tag must be done using dedicated hardware to reduce power consumption. Complex encryption, such as RSA and AES, while providing strong protection, currently requires too much power, silicon chip area, and time to be feasible for implementation into a passive RFID tag.

The power and area constraints on an active RFID tag are not as great as on a passive RFID tag. Therefore, an active RFID tag is capable of performing complex encryption, but this is not widely used. Complex power hungry encryption schemes must be used sparingly as the extra energy required to perform the encryption significantly reduces the lifetime of the active RFID tag. Further, encrypting a tag ID does not prevent tracking because a malicious user can simply track the encrypted tag ID.

The ISO 18000-7 standard provides the ability to lock the RFID tag. When locked, an ISO 18000-7 active RFID tag will not respond to some commands, but the memory can still be read using the collection with data command [7]. The goal of the password protection is to prevent changing existing or writing new data or operational parameters of the tag. The password is 32 bits and based on the time required to issue the command to disengage password protection. Brute force attempts to crack the password require about 124 days to try all possible passwords.

It is doubtful that a malicious individual would have access to the same tag for 124 days.

Securing RFID tags against clandestine reading is a difficult problem because of the widespread deployment of RFID tags. Protection against tracking is a difficult problem because many different legitimate entities must be able to read RFID tags. Common solutions, such as public/private keybased encryption, break down as any interrogator wishing to read a tag would need to obtain the key set from a central server. The latency of the key lookup could reduce read rate and provide difficulty for systems that do not have a connection to the central server or Internet. Methods, such as the kill command in Gen-2 tags (passive RFID tags), provide an extreme solution to this problem, but once killed the RFID tag can no longer provide any benefit to the owner. Because active RFID tags can operate without being interrogated, a second password could be used to prevent the active RFID tag from responding for a certain length of time. Because the active RFID tag is battery powered, time can be tracked and it could reactivate after the time period expires. This protection would provide limited protection against tracking (protected during the time interval) while allowing the benefits of the RFID tag to be available in the future (the kill command does not allow this). As the ISO 18000-7 system already requires management of one password, a second password can be added with very little difficulty.

Another approach is to safeguard the linkage between the tag ID number and the actual information about that tag ID number. A centralized database system containing information, for example, product identification or cargo lists, would be able to limit access to only legitimate users. EPCIS, under development by EPCglobal and the autoID labs, is one instance of such a system. This system would provide the clandestine listener with only the tag ID, or Electronic Product Code (EPC) number in the case of Electronic Product Code Information Services (EPCIS). The clandestine listener would then require access to the centralized database to obtain more information from the tag ID. Tracking is still possible with this system, but additional privacy protection is added. Currently this type of system is being set up as part of the larger RFID system to allow information to be used by the interrogator (i.e., to look up the price of an item) so it requires very little or no additional effort to deploy.

1.3.5 Increasing Battery Life

Because the onboard power supply of the vast majority of active RFID tags is a battery, increasing the battery lifetime is a critical issue. A simple inspection of the energy required for each of the primary tasks of an active RFID tag—processing, receiving, and transmitting—usually show transmitting as being the main energy consumer. While it is often true that to transmit one bit requires more power than to process or receive one bit, minimizing the transmitter energy usually does not yield much gain. This is because the receiver must always be listening for a message as the tag does not know when a message will arrive. Some active RFID tags can enter a low power sleep mode, but the receiver must still be active to listen for the wake-up command. Thus, over the lifetime of the active RFID tag, it is the receiver that consumes the majority of the energy. Therefore, minimizing the energy consumed by the receiver will provide the greatest increase in lifetime. This is important because during transportation it may be several days, or weeks between reads (e.g., when on a shipping container onboard a ship).

One method to reduce the receiver power to nearly zero is the burst switch developed at the University of Pittsburgh RFID Center of Excellence. The burst switch is a passive receiver coupled with an ultra-low power processor capable of detecting the presence of certain forms of energy [8]. The ultra-low power processor is used to decode the presence of energy to check if it resembles the wake-up code. Once the wake-up code is detected, the rest of the active tag can be wakened. The burst switch is effective in filtering out ambient RF noise and a sufficient wake-up code can prevent the active RFID from waking up in response to noise. With the burst switch, an active RFID tag can remain asleep the majority of the time, thus saving power and extending battery lifetime. The nanoTag, developed using the burst switch by the University of Pittsburgh RFID Center of Excellence, completely removes the active receiver producing a hybrid passive-active RFID tag, or a semiactive RFID tag [8].

1.3.6 Current Outstanding Problems with Active RFID Tags

The three major research areas that need to be addressed in the development of active RFID tags are: (1) low power communication hardware, (2) devices with low leakage current when dormant, and (3) enhanced security.

1.3.6.1 Low Power Communication

Low power communication hardware is critical as transmission and reception are the leading consumers of energy in an active RFID tag. Current communication hardware can reduce energy consumption by reducing the effectiveness of the receiver or transmitter leading to a reduction in the read range of the active RFID tag. As advances in more energy-efficient circuit design and fabrication processes for communication hardware is achieved, energy savings are possible. Advances of alternative methods, such as the burst switch, are also necessary.

1.3.6.2 Lowering Energy Consumption When Dormant

Devices with low leakage current are critical to the lifetime of an active RFID tag. Because active RFID tags spend most of their time in the sleep mode, minimizing the leakage current will minimize the energy consumed while asleep.

1.3.6.3 Enhanced Security

Security of the data contained in the RFID tag is critical in some applications. Further, securing the communications between the interrogator and tag is also important. Methods to protect data on the tag and to enable tags to identify and respond to only friendly interrogators must be developed. The latter problem is difficult if RFID is to be widely deployed, as each tag would need to know in advance which interrogators are friendly.

1.4 Semipassive RFID Tags

Semipassive tags use an onboard power supply to power the controller or microchip and may contain additional devices, such as sensors and communicate using backscatter. Semipassive tags are often used in environments that are too RF unfriendly for passive tags or where sensor readings must be taken periodically to monitor goods. While active tags can be used in both cases, active tags generally cost more than semipassive tags. Likewise, semipassive tags cost more than passive tags.

Both semipassive and active tags suffer from the problem governed by the battery lifetime. Because of their inexpensive cost relative to active tags, semipassive tags are often used although not reused, where an active tag may be reused many times.

Cold chain is a major area where temperature must be recorded at regular intervals. The term *cold chain* refers to the supply chain of any goods that must be kept within specific temperature limits, usually below freezing. Most often this refers to frozen foods and medications. Cold chain goods that are not kept within the required temperature range can be damaged. For example, food can spoil causing illness, and medications could alter their properties becoming useless or dangerous. Attaching a semipassive tag to each pallet enables the receiver of the shipment to determine if that pallet was kept at the appropriate temperature during transport. With this information, the receiver can reject and dispose of any goods that were not kept within the temperature limits.

1.4.1 Extending Read Range

A passive tag uses the energy contained in the interrogator signal for two purposes. First, the passive tag uses the part of that energy to power itself by absorbing the energy. Hence, the amount of energy that can be used for backscatter communication with the interrogator is reduced.

Two criteria must be met for an interrogator to communicate successfully with a passive tag. First, the passive tag must receive enough energy from the interrogator to turn on the integrated circuit (IC) components of the tag. As mentioned above, the powering of the IC components reduces the available energy for communication. Second, there must be enough energy left over for the tag to transmit the reply to the interrogator such that the interrogator is able to receive, detect, and decode the backscatter reply of the tag.

In a semipassive tag, the IC components are powered by a battery, thus allowing the semipassive tag to backscatter all the received energy from the interrogator, resulting in a stronger communication signal from the semipassive tag to the interrogator. The stronger communication signal from the tag to the interrogator increases the read range of a semipassive tag. Generally, semipassive tags can be read farther away from the interrogator or in a more RF unfriendly environment than a passive tag.

1.4.2 Equipping with Sensors

Semipassive tags can be equipped with sensors. While passive tags also can be equipped with sensors, the sensor cannot be powered in the absence of an interrogator signal. Hence, sensor technology requiring long warm-up times, often measured in hours, are not feasible for a passive tag. Electrochemical sensors, which generate either a voltage or current that is proportional to a specific chemical often require several hours to warmup and provide accurate readings when first powering up. Simple sensors, such as a thermistor, which is simply a variable resistor whose resistance changes when the temperature is known, do not require a long warm-up time. In many applications, it is infeasible to have an interrogator read the passive tag every time a measurement is required. Further, it may not be feasible or economical to place an interrogator at every point in the chain of custody. Finally, the inability to ensure a successful tag read 100 percent of the time can result in an incomplete data log.

Semipassive tags do not need an interrogator to supply power for a measurement. This enables sensors with a warm-up time to remain powered and allows readings to be taken by both types of sensors at the desired intervals. As long as the battery has enough energy remaining and the sensor is functional, the semipassive tag can guarantee a reading. Thus, with a durable sensor and appropriate battery, semipassive tags can guarantee a complete record of readings. Further, no provision must be made to provide an interrogator to accompany the goods being monitored and an unfriendly RF environment will not prevent sensor readings.

1.4.3 Outstanding Issues with Semipassive Tags

Three major areas of research are needed to improve semipassive tags. First, cost is a major issue to tagging at lower levels: case and item level. Second, lower power sensors are required to increase the lifetime of a semipassive tag. Third, development of a semipassive tag that could still function as an entirely passive tag would be very useful.

1.4.3.1 Cost

Temperatures can vary significantly within a container and this introduces the possibility of an acceptable temperature in one location, but an unacceptable temperature in another nearby location. If few tags are used, such as one per pallet, if that single tag is in the acceptable temperature area, there is no guarantee that the majority of that pallet is within the temperature limits. Case-level tagging will reduce this uncertainly and item-level tagging will provide even greater reduction in uncertainty and increase the protection offered by semipassive tags. However, the consumer, the product manufacturer, or the retailer must absorb the cost of the tag, and there is a limit to the amount of money each party will absorb. Lowering the cost per tag enables more goods to be monitored, as the cost that must be absorbed is within each parties limit.

1.4.3.2   Lower Power Sensors

The battery powering the sensors on a semipassive tag has a finite lifetime. Lowering the energy consumed by the sensors and IC components will increase the lifetime of the tag. Tag lifetime is critical because, to provide a complete record, the tag must be actively taking sensor readings during the entire time the goods that it is monitoring are in transit. Extending the lifetime of the tag will allow for complete monitoring over longer transit times. Shorter transit times will also benefit as the period between sensor readings can be reduced providing more detailed information about the goods.

1.4.3.3   Passive Operation as a Fallback

A semipassive tag that would still function, even in a reduced capacity, as a purely passive tag after the battery dies would be extremely helpful. With such a tag, the sensor readings up to the point where the battery died could still be retrieved. That information may be enough to verify a shipment was kept within the proper conditions.

1.5   Future of RFID

RFID has the potential to improve numerous processes and provide added safety to the public. RFID simply provides information that can be used to track and monitor goods, assets, and even people. This is of great advantage in the retail and food sectors to track goods, and in the pharmaceutical and medical sectors to track and verify the authenticity of drugs. Equipped with sensors, RFID tags can provide warnings when goods or medications are not kept within acceptable environmental limits. In the retail sectors, companies can use RFID to prevent the loss of goods in the supply chain and in the stores.

However, this ability can be used maliciously as RFID tags can be read remotely or the communication between the tag and interrogator can be observed. Using this information, a person could track or monitor another person and conjecture information about this person. For example, if drugs were tagged with RFID tags, it would be possible to determine what drugs a patient had obtained and then use that knowledge to access that person's medical condition. Such a release of private information could be harmful or unsettling to the violated person. As a result, security measures are needed before RFID becomes widely accepted. Alternative methods to killing the tag need to be researched to allow the tag to function, but at a reduced capacity. One example of limiting the effective read range of the tag is a viable solution [9]. More research is needed to develop methods to enable the information stored in the RFID tag to be used after the point of sale while reducing the amount of risk associated with malicious reading of the tag.

RFID is gaining ground, albeit slowly, in the retail sector, but a ROI in the general case has not yet been established. Without a clear ROI, few companies will adopt RFID technology. This is a similar case as with barcodes, which took several years to become as widespread as they are today [10]. For widespread adoption, one communications protocol must be established to allow interoperability. This is similar to the use of the OSI layering structures that enable the Internet access through a number of devices. It is not necessary for all RFID applications to use the same communication protocol. However, the communication protocol should be consistent throughout a particular application.

As RFID consists of LF, HF, UHF, and UWB (Ultra-Wideband) tags, each having different characteristics, advantages, and disadvantages, the choice of communications protocol will be influenced by the technology chosen and the application. What must be avoided is to have multiple systems in a single application area that cannot communicate with each other. In this case, all parties must purchase all systems that are used, or select a small number of different RFID products. Such an expense inhibits adoption or limits the number of people with whom business can be done without adding additional RFID infrastructures.

Sensor-equipped tags normally require a battery to power the sensor when the tag is not being interrogated. Battery life is critical to the usefulness of sensor-equipped tags because the tag must operate for the entire time the product is in transit. Hence, a longer battery life enables this protection to be extended to longer transit times. For goods traveling across oceans on ships, this is important, as the transit times could be several days to a few weeks.

RFID allows users to gather the information they need and want. However, RFID can also be used for malicious purposes to track and obtain private information about people. Therefore, RFID must be used appropriately and steps must be taken to inform people about the presence of RFID and to provide safeguards and countermeasures against illicit use of RFID. Further, as RFID encompasses many different technologies, RFID education is critical to enable users to select the RFID system that best meets their needs. When used appropriately, RFID offers great promise to the world community from reducing manufacturing and retail costs to safeguarding human lives. When employing RFID, openness and transparency about the use of RFID is probably the best way to proceed.

References

[1] D. Halliday, R. Resnick, and J. Walker. *Fundamentals of Physics*, 5th ed. New York: John Wiley & Sons, 1997.

[2] J. Landt. "The History of RFID." IEEE Potentials, 24, 4: 8–11, Oct.–Nov. 2005.

[3] C. Swedberg. "Macau Casinos Use RFID to Authenticate Chips." *RFID Journal*, Dec. 7, 2006, Internet: http://www.rfidjournal.com/article/articleview/2878/ [accessed Mar. 27, 2007].

[4] K. Finkenzeller. *RFID Handbook: Fundamentals and Applications in Contactless Smart Cards and Identification*, Hoboken, NJ: John Wiley & Sons, 2003.

[5] U.S. Department of Commerce, National Telecommunications and Information Administration. *Manual of Regulations and Procedures for Federal Radio Frequency Management*. May 2006.

[6] Alien Technology. "EPCglobal Class 1 Gen-2 RFID Specification," 2005. Internet: http://www.alientechnology.com/docs/AT_wp_EPCGlobal_WEB.pdf [accessed Feb. 13, 2007].

[7] ISO, Information Technology Radio-Frequency Identification for Item Management, Part 7: Parameters for Active Air Interface Communications at 433 MHz, ISO/IEC FDIS 18000-7, 2004.

[8] P.J. Hawrylak, L. Mats, J.T. Cain, A.K. Jones, S. Tung, and M.H. Mickle. "Ultra Low-Power Computing Systems for Wireless Devices." *International Review on Computers and Software*, 1, 1: 1–10, July 2006.

[9] M.C. O'Connor. "IBM Proposes Privacy-Protecting Tag." *RFID Journal*, Nov. 7, 2005, Internet: http://www.rfidjournal.com/article/articleview/1972/1/1/ [accessed Mar. 27, 2007].

[10] S. Baron and A. Lock. "The Challenges of Scanner Data." *The Journal of the Operational Research Society*, 46, 1: 50–61, Jan. 1995.

[11] M.C. O'Connor. "Gillette Fuses RFID with Product Launch." *RFID Journal*, Mar. 27, 2006, Internet: http://www.rfidjournal.com/article/articleview/2222/ [accessed Mar. 27, 2007].

Chapter 2

RFID Automatic Identification and Data Capture

Xiaoyong Su, Chi-Cheng Chu,
B.S. Prabhu, and Rajit Gadh

Contents

Automatic identification data capture (AIDC) technologies are becoming increasingly important in the management of supply chain, manufacturing flow management, mobile asset tracking, inventory management, warehousing, and any application where physical items move through location on time. Tracking these items has historically been done by the use of bar-code technologies, which suffer from lack of efficiency, robustness, difficulty in automation, inability to have secure or dynamic data, etc., whereas the electronic technology of Radio Frequency IDentification (RFID) has the ability to overcome several of these barcode limitations This paper presents a comparative basis for the creation of on AIDC infrastructure via RFID versus other technologies, such as barcode and sensor technologies.

2.1 Introduction

As the communication and computational technologies have started to become commonplace within enterprise operations, the computer-aided/managed Enterprise Information Systems (EIS), such as Enterprise Resource Planning (ERP) [1], Supply Chain Management (SCM) [2][3], Product Lifecycle Management (PLM) [4][5], Customer Relationship Management (CRM) [6][7], Manufacturing Execution System (MES) [8], Warehouse Management System (WMS) [9], and Enterprise Asset Management (EAM) [10], are significantly improving the enterprise operational efficiency and reducing the operational cost [11]. The EIS processes information, such as history, current status, location, relationships, and destination of enterprise resources, such as materials, equipments, personnel, cash, etc.

While the EIS manages all types of resources, physical enterprise resources are the object of interest in the current research—they include materials, equipments, personnel, and products, and they are defined as "business objects." Usually, business objects are mobile and subject to constant modification within or across the enterprise boundary. Among other capabilities, the EIS must track, trace, locate, predict, and work on these objects in real-time to improve efficiency and productivity of enterprise operations. For example, the stock level of products at various stages within a supply chain can significantly affect the operations of the supply chain. Optimum stock levels result in requiring less storage space, faster processing, quicker cash flow, better customer satisfaction, and sales. The stock level can be monitored and predicted by tracking the movement of the on-shelf products, or the incoming shipments. As the real-time stock levels are identified, decisions, such as replenish or reorder can be made timely and correctly. This is called visibility and predictability in the enterprise information system.

To achieve visibility and predictability of the movement of these business objects, the information associated with the object, which is called identification data in this research, should be identified and monitored along the enterprise operation flows. The identification data should automatically be captured and integrated into the different enterprise process applications in real-time. Usually, the identification data capture process and the integration of the identification data with enterprise application are performed by the AIDC technologies. The identification data associated with a particular business object (such as raw material, products, equipments, shipments, and personnel) is collected by the data capture devices at each location where the business object is processed. The barcode is the most commonly used identification data capture technology in today's enterprise operations. However, traditional barcoding approach cannot achieve the real-time visibility because of the low speed of reading, the needs of line-of-sight, and unavoidable involvement of humans. The more advanced AIDC technology of RFID is becoming the promising technology to achieve real-time visibility of enterprise operations. It has several obvious advantages, such as nonline-of-sight reading, high-speed reading, multiple reading and writing simultaneously, minimal human intervention, etc., that make it close to ideal for providing real-time visibility of enterprise operations.

2.2 Identification Automation Technologies

Barcode, RFID, sensor, magnetic strip, IC card, optic character recognition (OCR), voice recognition, fingerprint, and optical strip [12] are identification technologies that are being used in the enterprise environment. Among these identification technologies, barcoding is the most widely used technology. However, the RFID and sensor hold the most promise of significantly improving business operational efficiencies and increasing the visibility of the business objects. The other technologies have either a lack of automation capability or a lack of ability to attach to business objects. Thus, we do not categorize them as the automatic identification technology for enterprise application. Barcode, RFID, and sensor are the three technologies addressed and discussed in this chapter.

2.2.1 Barcode Technology

2.2.1.1 History

The first barcode was developed by Bernard Silver and Norman Joseph Woodland in the late 1940s and early 1950s [13]. It was a "bull's eye" symbol that consisted of a series of concentric circles. The first commercial use of barcodes was by the RCA/Kroger system installed in Cincinnati on the behest of the National Association of Food Chains (NAFC). However, it was not widely used until the Universal Product Code (UPC) [14] was introduced in America and adopted by the U.S. Supermarket Ad Hoc Committee. Today's barcodes have two forms: one dimensional (1D) barcode and two dimensional (2D) barcode. The 1D barcodes use bars and gaps to encode identification information such as serial numbers. The 2D barcodes consist of more complicated patterns and may encode up to 4 kilobytes of data. Figure 2.1 shows the

' % DUFRGH   '%DUFHG

Figure 2.1   Two different types of barcodes.

___

two types of barcodes. Although 1D is the more prevalent barcode used in daily life, the 2D barcode is becoming increasingly popular because it needs significantly lower surface area to encode the same amount of data as compared to 1D barcodes.

Barcodes can be printed from most printers. One dimensional barcodes usually have coded readable ID printed along with the barcode. Barcodes can be read by barcode scanners, which we see at a typical point of sale (POS) in retail stores.

Figure 2.2 illustrates a basic barcode system. Barcodes are read or scanned by a barcode reader and the reader is connected to a computer. The operator has to physically align or point the barcode reader with or to the barcode to read the identification information. The computer software processes the identification information picked up by the scanner. Programmable logic controller (PLC) is often used to control the scanner in a more automated process, such as the production line. The primary scanning technology for barcode is LED (light-emitting diode). More advanced scanning, such as CCD (charge-coupled device), laser, and imager are used in industry automatic processing [15].

Figure 2.2   A basic barcode system.

2.2.1.2   Symbology

The coding scheme that barcodes use to encode data is called symbology [16]. A symbology defines how to encode and decode the barcode data. Some symbologies can encode numbers only and others can encode both numbers and alpha letters. Each symbology normally is designed for certain applications, but it could extend beyond that. A 1D barcode usually contains only an identification number while a 2D barcode may contain customized data. Table 2.1 shows commonly used symbologies and their applications [16].

2.2.1.3   The Advantages and Disadvantages of Barcodes

Compared to manual data entry, the barcode is fast and accurate. The barcode can be printed from any black/white printer. Since the barcode can be directly printed on an

Table 2.1   Commonly Used Barcode Symbologies

	Symbology	Notes	Applications
1D	CODE39	Letters, digits, and a few special characters	Processing industry, logistics, library, manufacturing, military and DoD, healthcare
	CODE93	Can encode all 128 ASCII characters; it's the enhancement of CODE39	Same as CODE39
	CODE128	Besides the 128 ASCII characters, it can encode four special function codes	Shipping, logistics
	EAN/UCC 128	Similar to CODE128	Shipping, logistics
	CODABAR	It can encode characters: 0123456789-$:/.+ABCD; ABCD are start and stop characters	Library, blood bank, air parcel business, medical/ clinical applications
	CODE 2 of 5	Digits only, high density; includes three variants: standard, interleaved, and industrial	Automotive industry, good storage, shipping, heavy industry
	UPC A and UPC E	Digits only, UPC A is 12 digits and UPC E is 7 digits	Retail, consumer products
	EAN13 and EAN8	Europe version of UPC	Retail, consumer products
	RSS	Latest barcode symbology; it is used for space constrained identification; It has different variants and could be composted with other symbologies	Use only when necessary
2D	PDF417	High-density 2D barcode can store up to 2725 characters	Shipping, defense, automotive
	DATA MATRIX	High density can store up to 3116 characters	Defense, automotive
	MAXICODE	A fixed size matrix has up to 93 alphanumeric or 138 number characters; supports high-speed scan and is orientation independent	Shipping
	AZTEC CODE	High-density 2D barcode can encode up to 3750 characters	Retail, assets, consumer products

object or on paper labels, the cost for a barcode is typically less than one penny [17]. Even after including the hardware cost, the barcode data collection system reduces the operation cost, labor cost, and the revenue loss caused by data entry errors, while improving the business process and productivity [18]. However, several weaknesses exist. First, a barcode label is easy to damage in harsh environments, such as careless handling, external factors, such as rain and low temperatures. Second, to read the barcode, the barcode scanner needs to be in the line of sight with the label. It means that the manual movement of the objects or scanner is necessary. Third, barcode technology does not have the ability to scan an object inside a container or a case. Thus, the operator has to open the container and scan an the objects one by one, thereby involving intensive labor. Obviously, the barcode is incapable of fast processing.

2.2.2 RFID Technologies

2.2.2.1 History

The use of RFID can be traced back to World War II. A transmitter that can return proper response to an interrogator was attached to allied aircraft to identify friendly aircraft from enemy aircraft. From the 1960s to 1980s, RFID was used in fields, such as hazardous material processing and livestock tracking [19]. Then, during the Gulf War, the RFID technology was successfully used for transporting military goods to the battlefield. A new generation of RFID surged when the retail giant Wal-Mart mandated its major suppliers to ship their goods to the Wal-Mart distribution centers with enabled RFID tags. Similar mandates then originated from the Department of Defense (DoD) and other retailers, such as Target, Metro, Albertson's, etc. RFID technology is broadening to other industries, as well including healthcare, manufacturing, life science, transportation and logistics, banking, and others. Moreover, RFID is moving trading partners toward cross enterprise collaboration.

2.2.2.2 Existing RFID Technologies

RFID technologies can be classified into three categories: passive RFID, active RFID, and semipassive RFID [20]. Based on the radio frequency used, the passive RFID technologies are usually categorized into low frequency (LF) RFID, high frequency (HF) RFID, ultra high frequency (UHF) RFID, and microwave RFID. The essential components of RFID hardware are:

1. **RFID tag:** A tiny silicon chip attached to a small antenna.
2. **Reader antenna:** Used to radiate energy and then capture the energy sent back from the tag. It can be integrated with the reader or connected to the reader by cable.
3. **Reader:** The device that talks with tags. A reader may support one or more antennae.

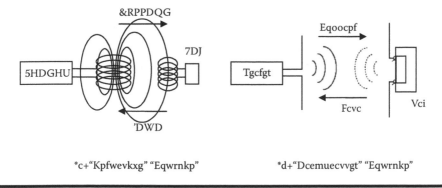

Figure 2.3   RFID coupling and operation.

The passive RFID tag powers up and exchanges commands/responses by gathering energy from the RF transmitted from the reader antenna by means of inductive coupling (LF and HF) or backscatter coupling (UHF). (Figure 2.3a illustrates the inductive coupling and Figure 2.3b illustrates the backscatter coupling.) The inductive coupling uses the magnetic field while the backscatter coupling uses electromagnetic waves to exchange data between the reader and the tag [20]. Figure 2.4 is a list of form factors of RFID tags, antennae and readers.

Table 2.2 lists the properties of passive RFID technologies. The sense (reading and writing) range of RFID technology depends on the antenna size of both tag and reader, the power level of the radio wave, sensitivity of tag and reader, tagged object, and environment. Frequencies and power used by UHF RFID technologies are regulated, and are different from region to region [21]. For example, the allowed frequency is 902 to 928 MHz in America, while the allowed frequency in Europe is 865.6 to 867.6 MHz. The power levels in each region are also different. For example, 2W ERP (effective radiated power) is allowed in Europe while 4W EIRP (Equivalent Isolated Radiated Power) is allowed in America.

An active RFID tag uses an onboard battery to support reader–tag communication. Currently, there is a lack of standards on active RFID technology. Most active RFID technologies are proprietary, and typical active RFID operates at 333 MHz

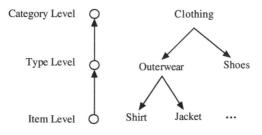

Figure 2.4   RFID hardware form factors.

Table 2.2   Passive RFID Technologies

	LF	*HF*	*UHF*	*Microwave*
Frequency	Typically 125 or 134.2 KHz	Typically 13.56 MHz	Typically 868 MHz–928 MHz	2.45 GHz and 5.8 GHz
Reading/ writing data rate	Slower	Moderate	Fast	Faster
Ability to read/write multiple tags in field	No tag will be read if multiple tag present	Can read/write moderate dense tags	Ability to read/write high-density tags	Similar to UHF
Approximate range	Up to ~1 m	Up to ~1.2 m	Up to ~6 m	Up to ~2 m
Limitations (cannot work on)	Metal	Metal	Metal, liquid	Metal, liquid
Applications	Animal tracking, access control vehicle immobilization	Smart card, access control, item level tagging (supply chain, asset tracking, library application, pharmacy application, personnel identification etc.)	Pallet and case tagging (supply chain); item level tagging is working in progress	Similar to UHF

or 433 MHz. Wi-Fi-based active RFID works at 2.4 GHz. Compared to passive RFID, the active RFID usually has a longer reading and writing range (up to several hundred meters), bigger data storage space (up to a few megabytes) and faster reading and writing speed. The active RFID operations are less affected by environmental factors. Thus, active RFID is widely being used for tracking and locating large objects, such as containers and trucks. It can also be used for tracking livestock.

Like the active tag, the semipassive (semiactive) tag also contains a battery. However, the battery is usually used to assist in collecting environmental parameters, such as temperature or humidity. The communication between the reader and tag still uses the RF energy transferred from the reader and has the same characteristic as passive RFID.

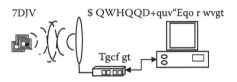

Figure 2.5   A basic RFID system.

2.2.2.3   A Basic RFID System

A combination of RFID technology and computing technology that brings value to a business or engineering process is called a RFID system. The simplest RFID system contains tags, a RFID reader, and a host computer, as shown in Figure 2.5.

- **Tag**: Depending on its final form factor, the tag can be categorized into labels, transponders and inlays, as shown in Figure 2.6. To simplify the description, the term *Tag* is being used to represent all three types of tags. The tag is attached to an object.
- **Reader:** A reader has one or more antennae that are either external or internal. The interfaces that connect the reader with the host computer are one or more of the following: RS232, RS485, USB, Ethernet, Wi-Fi, PCMCIA, or CompactFlash. The reader reads the data ID of a tag through the antenna. The data is then sent to the host computer for further processing. Traditional RFID readers only have the function of capturing data.
- **Host computer:** Controls the reader behavior and dispatches the data captured by the reader to the carrect application.

2.2.2.4   Standards

To achieve interoperability of RFID devices and exchangeability of RFID information, various standards or protocols are proposed for different applications [22]. These standards include hardware physics specifications, tag–reader air interface specifications, reader–host command specifications, reader network standards, and data formats.

Transponder   Label   Inlay

Figure 2.6   Tags.

International Standards Organizations (ISO) 18000 series of standards define RFID air interface operations, which include physical layer electronic characteristics and data link layer for data communication. ISO 18000 contains seven parts:

Part 1 defines general parameters.
Part 2 specifies the parameters for frequency under 125 KHz (LF).
Part 3 specifies parameters for 13.56 MHz (HF).
Part 3 and Part 4 work on 2.45 GHz and 5.8 GHz, respectively.
Part 6 works on frequency between 860 MHz to 930 MHz (UHF).
Part 7 works on 433 MHz (used for active RFID).

Besides the ISO 18000 series of standards, ISO 14443 and ISO 15693 define the air interface and communication protocols for the proximity system and vicinity card. ISO 14223 defines the air interface for identification of animals in agricultural applications.

Along with the air interface definitions, there are additional standards for rectifying the processes of tagging, packaging, data coding, and other special applications. For example, ISO 15961-15963 series define the concept, data syntax, business process, and information exchange for item level management; ISO 17363-17368 define the different types of packaging processes for the logistic and supply chain. ISO 11784 and ISO 11785 define the code structure and application concept for animal identification. For each RFID technology there is a conformance test standard to guide its use and test.

The Electronic Product Code (EPC) [23] was originally developed by the Auto ID Center at the Massachusetts Institute of Technology (MIT). In 2003, the Auto ID Center licensed the EPC technologies to EPCglobal, a joint venture between European Article Number (EAN) International and the Uniform Code Council, Inc. Since then, EPCglobal has taken the mission of standardizing and developing hardware physics, communication protocols, the EPC network infrastructure specifications, the supported software specifications, and data formats.

EPCglobal defines the EPC classes according to the functional characteristics. The classes are [24]:

■ Class 0: "Read Only" passive tags. The ID is programmed by the manufacturer.
■ Class 1: "Write-Once, Read-Many" passive tags. The ID is programmed by the user and locked to prevent rewrite.
■ Class 1 Generation-2 (Gen-2): An updated version of the Generation-1. It supports more secure operations and contains optional user memory.
■ Class 2: Multiple Read/Write passive tags. It contains user memory that allows updating and encrypting.
■ Class 3: Multiple Read/Write semipassive tags with user memory.
■ Class 4: Active tags.
■ Class 5: Readers.

The information exchange of the EPC complaint application is based on the data structure or coding scheme defined in the EPC standards. The first generation of the

EPC coding scheme is quite simple. The 96 bits or 64 bits of a RFID tag number has been divided into four segments, which include: Header, EPC Manager, Object Class, and Serial Number. However, this scheme cannot represent the current global trading business-related identification numbers. The latest EPC coding scheme supports more coding that is being used in current global trading business. These coding schemes include: Serialized Global Trade Identification Number (SGTIN), Serial Shipping Container Code (SSCC), Serilized Global Location Number (GLN), Global Returnable Asset Identifier (GRAI), Global Individual Asset Identifier (GIAI), and General Identifier (GID). The identification coding scheme for the U.S. Department of Defense (DoD) are also supported in the EPCglobal EPC tag standard document [25]. The EPC Gen-2 air interface protocol for UHF communication has been proved as ISO 18000-6C [26].

2.2.2.5 Advantages and Disadvantages

Compared with other identification technologies, RFID has the following advantages:

1. High reading speed, multiple reading and writing simultaneously: Depending on the technology used, the reading speed of RFID can be up to 1000 tags/sec [23]. Higher reading speed leads to higher throughput in the system. For example, in a high-speed sorting system, the throughput could be improved significantly by improving the speed of identifying an item, which could be the bottleneck of the system. Inventory management would be improved by using RFID because the item-by-item manual checking can be substituted with an automated method.
2. The RFID technology does not need to be line-of-sight. For example, when shipping a container with RFID tags installed on products within the container, it is not necessary to open the container and read the tagged items one-by-one because the readers can typically read tag data through or around blocking materials.
3. The data carried by an RFID tag is rewritable. The memory storage ranges from several bytes to a few megabytes. With the user memory, the business information carried by the tag can be changed during any point of the process dynamically. More data can be carried on the tag itself so that the tightly coupled back end database may not be required, which is not the case in the traditional AIDC approach.
4. RFID has a longer reading/writing range compared to barcodes and most other identification technologies. The sensing range of RFID tags varies from a few centimeters to hundreds of meters [20].
5. Security checks can be added into RFID tags to prevent unauthorized reading and writing. Fear of the disclosure of privacy slows down the practice of many advance technologies. The RFID is one of them. Fortunately, a more sophisticated authentication encryption algorithm is being implemented

in the RFID tags and readers. By adding such security check functions into the implementation, RFID is becoming the dependable technology to prevent ID theft and counterfeiting. On the contrary, its close counterpart—barcode—does not have such capability.

6. The combination of RFID and sensor technology could bring more value to enterprise applications. For example, during the food transportation, it is important that the temperature of the food be monitored and controlled. Normally the temperature is set on the refrigeration device. In a large container, the temperature may vary at different points within the containers, especially when large numbers of items are being transported. An RFID tag combined with a temperature sensor could locate and precisely monitor the status of each individual item.

Therefore, RFID is more effective when faster processing, longer read range, flexible data carrying capability and more secure transactions are required. Some of the widely acknowledged benefits of RFID are that it could:

1. Improve warehouse and distribution productivity.
2. Improve retail and point-of-sale productivity.
3. Reduce out-of-stock and shrinkage.
4. Help prevent the insertion of counterfeit products into supply chains. [27].

However, despite these obvious advantages, there exist some disadvantages as well:

1. A major drawback of RFID is its cost. The current price for each UHF tag is still costly. The RFID industry is trying hard to reduce the manufacturing cost. Within a few years, the industry hopes to reduce the cost to about 3 cents per tag for UHF tags. A 5 cent RFID tag is already available [28].
2. Reading and writing reliability are largely affected by the material of the tagged product and its surrounding environment. For example, it is difficult to read a UHF RFID tag surrounded by liquid or metal. By studying the physics and doing experiments, we observed that by properly placing or encapsulating the RFID tag, the reading performance was improved [29]. New methodologies are being developed to overcome the challenges faced by environmental factors. For example, new UHF tags and readers based on Near-Field Communication (NFC) developed by Impinj demonstrates that NFC UHF RFID tags can be read even when the tags are in liquid [30]. Various tag encapsulation methods and RF technologies are being researched to improve the performance of RFID. We believe that these problems will be solved by technological advancements in this field.

2.2.3 Sensor Technologies

The sensor is a device that can measure and collect environmental parameters (such as temperature, humidity, chemicals, vibration, density, etc.) or system run-time parameters (such as position, location, speed, acceleration, etc.) [31]. Sensors have been used in a variety of applications, especially for automation and control in industries, such as aerospace, automotive, healthcare, environment, transportation, etc. As the smallest unit of an automation system, a single sensor can be used for fulfilling certain functions. But, in most cases, a sensor system consists of a large number of sensors that must work together to achieve awareness of the physical surroundings.

Recently, new technologies, including wireless sensors, MEMS sensors, smart sensors, bio sensors, etc., have changed the type of sensors that can be made. Sensors made today with advanced techniques are smaller and can measure and collect information that is beyond the capability of traditional sensors. New sensor system infrastructures, such as sensor network [32] with wireless connection capability, are gradually replacing the traditional wire-connected sensors.

In an enterprise environment, sensor technologies combined with ID technologies would significantly improve the enterprise resource visibility and, thus, improve the enterprise operation efficiency. For example, along the supply chain flow, a global positioning system (GPS) sensor could help the enterprise system track and monitor the location of the raw material in real-time [33]. A better decision on the related enterprise operations could then be made based on information of not only on the location of an asset, but also its condition and, then possibly, the condition of the asset could alter where it was being next sent (for example, if icecream was being shipped at high temperatures, it could be disposed of instead of continuing to the final destination).

2.3 AIDC Infrastructure

The advantages of identification technologies can potentially improve the enterprise operation efficiency and reduce the operation cost. However, simply deploying data-capturing devices, assigning identification data to the business object, and capturing the data from the business object cannot alone bring value to enterprise operations. A set of software components, which assist devices management, identification data preparing, capturing, formatting, and associating with physical objects, are required. The software components and devices together are called identification resources. Further, these identification resources are networked and collaborate with each other to form the AIDC infrastructure.

2.3.1 Definitions and Components

The AIDC infrastructure is defined as a set of networked devices and software components, which include:

- **Devices:** Various identification technologies, such as RFID reader, RFID printer, barcode scanner, sensors, Programmable Logic Controller (PLC), etc.
- **Services:** Software components that enable the data preparation, capturing, and processing.

Essential components of an AIDC infrastructure are identified [34] [35] and illustrated in Figure 2.7. It contains the following components:

■ **Barcodes, Tags, and Sensors:** The smallest units that are attached to an enterprise entity or resource to be identified.

■ **Device Controller or Edge Server:** A device controller is used to manage and control identification hardware (readers, scanners, sensors and other manageable devices), aggregate, preprocess, and cache the identification information. It is difficult to manage a device in a planer space. Putting all the functions, such as capture, process, and representation of identification data, into a single server could increase the burden of the server. Hierarchical, layered, and distributed architecture is an effective way to balance the load and deliver the best overall performance. Using the edge servers, the devices can be clustered and distributed. Functional roles can be separated so that the edge servers can be dedicated to the data capture requirements where the core and central servers can be used for data intelligence.

■ **Identification Network:** The infrastructure that connects all of the hardware resources and enterprise information systems together.

■ **Enterprise Information Servers:** Provides enterprise activities-related data, which can be used along with the identification information for business operations. It provides real-time, aggregated identification data and

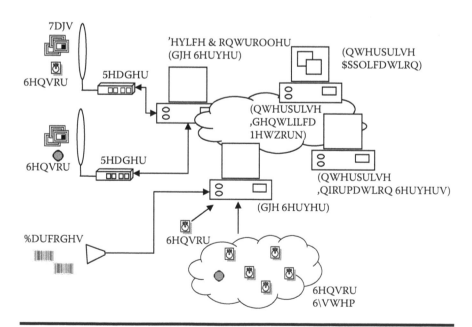

Figure 2.7 The AIDC infrastructure.

events to client applications. As discussed earlier, the identification data capture process may contain business events. The Enterprise Information System provides interfaces so that the application can define, register, and look up events. It also provides interfaces that the end application can register and look up production information, business information, and transaction information that is associated with a particular identification data.

■ **Enterprise Application:** Functional modules that fulfill certain enterprise activities. For example, a Warehouse Management System uses the data captured by the edge server to monitor the inventory level; an Asset Management System uses the data to look up a particular asset, etc.

Identification data carried by barcodes, RFID tags, and sensors is collected by a data capturing device and passed on to the edge servers or device controllers. It is processed at the edge server before being sent into the network. The components are not standalone—they are networked with other components and can exchange information with each other. For example, the identification data captured by one edge server can be used as the reference to generate an event by next edge server. If the number of IDs on the same pallet has been read differently by different edge servers, an event indicating that there is a missing product would be generated.

2.3.2 Challenges of Creating RFID-Oriented AIDC Infrastructures

As advanced AIDC technologies, such as RFID, are introduced, the challenges of creating AIDC infrastructures are present. The challenges are:

1. *The performance of data capture and processing*: High volume of data is required to be captured and processed at high speed. First the RFID system reads data 10 to 100 times faster than traditional barcode scanning [17]. Second, the ability to read at much greater distances and in larger region volumes around the reader results in greater volumes of data generated due to larger numbers of tags in the read-range of a reader. For example, if a carton has 100 tagged items, RFID reader is able to read all the items in less than one second because the UHF RFID reader has the ability to read up to 1000 reads/sec [23]. Thus, it is important that the AIDC infrastructure has the ability to capture such high volume data and rapidly send it to the end application. Providing real-time, accurate, and consistent data is the key to a streamlined enterprise operation. If the data is not accurate or is out of date, the wrong decision made based on the data would be costly to the enterprise. For example, consider the case of a supply chain in which the data captured during the transportation of goods cannot be sent to the ordering system on time. To control the inventory at a given level, the ordering system may request another order to fulfill the inventory control and this can result in duplicate orders.

2. *The manageability and extensibility of heterogeneous devices*: Co-existence of various identification devices makes them extremely difficult to manage, deploy, and configure. The diversity of RFID technologies (various standards, different frequencies, and utilizing passive and active technology) increases the difficulty of managing the data capturing devices. How to effectively manage devices and how to adopt new devices without significantly modifying the AIDC infrastructure needs to be researched.

3. *Scalability and performance of the AIDC infrastructure*: As high-volume data is being captured by the data capturing devices at high speed, how to ensure that the performance of the system would not degrade dramatically as the AIDC infrastructure grows can be a problem. For example, 7 terabyte data generated by the Wal-Mart RFID trial every day is reported in [36]. How to handle such high volume data in an AIDC infrastructure is challenging.

4. *Discoverability and availability of resources*: In a large scale enterprise environment, various services may be deployed in a logically and geometrically distributed environment. The client needs to find the right resource and obtain useful data from it. Either using the on-demand request (pull) or the real-time event (push) approach to provide the data and resource, the identification resources should be available and consumable to the client at any time.

5. *The integration of various identification resources with the diverse EIS*: The tasks of an AIDC infrastructure are not just to capture identification data. It acts as a bridge which it flawlessly connects all EIS together. However, the diversity of the data formats, especially when customized data is involved, increase the complexity of the data processing for the end application. The identification data should be associated with the physical object or business transaction so that it can be used in the EIS system.

2.4 Summary

A variety of identification technologies have been used in the enterprise systems to improve the operation efficiency and reduce the overall operation cost. Barcodes, RFID, and sensors are the most commonly used and important technologies that have been addressed. Because of its low cost, the barcode is the major identification technology used by most enterprises today. The emerging RFID technology brings significant new opportunities as well as challenges to the AIDC infrastructure. The combination of RFID and sensor technology adds additional value to the enterprise operations. For example, by combining the temperature sensor and RFID tag, the reader obtains the environment temperature of the tagged object. This is useful in cold chain where the temperature of the transporting product has to be precisely monitored and controlled. Although the RFID technology still has limitations, such as tag readability due to physical constraints, enterprises are increasingly using RFID to assist their business operations because of the advantages it brings.

Because of the emerging RFID technology and the diversity of identification devices, challenges exist in implementing an AIDC infrastructure. The challenges consist of the performance of data capture and processing, the manageability and extensibility of heterogeneous devices, scalability and performance of the AIDC infrastructure, discoverability and availability of resources, and the integration of various identification resources with the diverse EIS. Important topics of current AIDC research include the extensibility, scalability, resources usability and discoverability, RFID system performance, device manageability, and system integration ability of AIDC infrastructures.

References

[1] Christina Soh, Sia Siew Kien, and Joanne Tay-Yap, Enterprise resource planning: cultural fits and misfits: Is ERP a universal solution? *Communications of ACM*, 43, 47–51, 2000.
[2] Benita M. Beamon, Supply chain design and analysis: Models and methods, *International Journal of Production Economics*, 55, 281–294, 1998.
[3] D.M. Lambert and M.C. Cooper, Issues in supply chain management, *Industrial Marketing Management*, 29, 65–83, 2000.
[4] Valerie Thomas, Wolfgang Neckel, and Siguad Wagner, Information technology and product lifecycle management, In Proceeding of the 1999 IEEE International Symposium on Electronics and the Environment.
[5] Ioannis Komninos, "Production Life Cycle Management." http://www.urenio.org/tools/en/Product_Life_Cycle_Management.pdf (accessed...).
[6] M. Stone, N. Woodcock, and M. Wilson, Managing the change from marketing planning to customer relationship management, *International Journal of Strategic Management: Long Range Planning*, 29, 5: 675–683(9), October 1996.
[7] Russell S. Winer, "Customer Relationship Management: A Framework, Research Directions, and the Future." http://groups.haas.berkeley.edu/fcsuit/PDF-papers/CRMpaper.pdf, accessed November 2006.
[8] Ronelle Russell, Manufacturing execution systems: Moving to the next level, *Pharmaceutical Technology*, January 2004, 38–50.
[9] Ben Worthen, "The ABCs of Supply Chain Management." http://www.cio.com/research/scm/edit/012202_scm.html, accessed November 2006.
[10] IBM, "Enterprise Asset Management: Optimizing Business Operation with Leading-Edge Enterprise Asset Management Services." http://www935.ibm.com/services/us/index.wss/offering/bcs/a1002719, accessed November 2006.
[11] Diane M. Strong and Olga Volkoff, A roadmap for enterprise system implementation, *Computer*, June 2004, 22–29.
[12] Cambridge Consultant, "Description of Technologies and Example of Application in Other Industries." http://www.npsa.nhs.uk/site/media/documents/811_Cambridge ConsultantsReportSection 2.pdf, accessed November 2006.
[13] Mery Bellis, "Bar Codes." http://inventors.about.com/library/inventors/blbar_code.htm, accessed November 2006.

[14] GS1 US, "The Universal Product Code." http://www.uccouncil.org/upc_background.html, accessed November 2006.

[15] "Scanning technologies" http://www.mecsw.com/reviews/scanners/scantech.html, accessed November 2006.

[16] "Bar codes" http://www.lintech.org/comp-per/06BARCD.pdf, accessed November 2006.

[17] J. Abernethy, "Bar Codes & RFID, An Engineering Science 96 Report, M17," 1995. http://www.deas.harvard.edu/courses/es96/spring1995/barcodes_rfid/abernethy.html, accessed November 2006.

[18] Alexander Brewer, Nancy Sloan, and Thomas L. Landers, Intelligent tracking in manufacturing, *Journal of Intelligent Manufacturing*, 10, 245–250, 1999.

[19] Jerry Landt, "Shrouds of Time: The History of RFID." http://www.aimglobal.org/technologies/rfid/resources/shrouds_of_time.pdf, accessed November 2006.

[20] K. Finkenzeller and G. Munich, Radio-frequency identification: Fundamentals and application, *RFID Handbook*, John Wiley & Sons, New York, 1999.

[21] EPCglobal, "Regulatory status for using RFID in the UHF spectrum, Nov. 24, 2006." http://www.epcglobalinc.org/tech/freq_reg/RFID_at_UHF_Regulations_20061124.pdf, accessed December 2006.

[22] Craig K. Harmon, "RFID Standards." http://www.autoid.org/2002_Documents/hottopics/Standards_20020324.ppt.

[23] EPCGlobal, "Standards." http://www.epcglobalinc.org/standards, accessed December 2006.

[24] Impinj, "RFID Standards." http://www.impinj.com/page.cfm?ID=aboutRFID Standards, accessed December 2006.

[25] EPCglobal, "EPC Generation 1 Tag Data Standards Version 1.1 Rev. 1.27, Standard Specification, 10 May 2005." http://www.epcglobalinc.org/standards, accessed December 2006.

[26] Mary C. O'Connor, Gen 2 EPC protocol approved as ISO 18000-6C, *RFID Journal*. http://www.rfidjournal.com/article/articleview/2481/1/1, accessed December 2006.

[27] Tom Pisello, The ROI of RFID in the supply chain, *RFID Journal*, August, 2006. http://www.rfidjournal.com/article/articleprint/2602/-1/1, accessed November 2006.

[28] Mark Roberti, A 5-cent breakthrough, *RFID Journal*. http://www.rfidjournal.com/article/articleview/2295, accessed November 2006.

[29] Tadej Semenic, RFID radio study, RFID System Course project report, WIN-MEC, UCLA, Fall 2005.

[30] Impinj, "Impinj GrandPrix In Action." http://www.impinj.com/page.cfm?ID=RFIDVideos, accessed January 2007.

[31] Ljubisa Ristic, Sensor Technology and Devices, Artech House Publishers, Norwood, MA, 1994.

[32] Jason Hill, Robert Szewczyk, Alec Woo, Seth Hollar, David Culler, and Kristofer Pister, System architecture directions for networked sensors, in Proceedings of the 9th International Conference on Architectural Support for Programming Languages and Operating Systems, ACM, Cambridge, MA, Nov. 2000, 93–104.

[33] Harish Ramamurthy, Dhananjay Lal, B.S. Prabhu, and Rajit Gadh, Re,WINS: A distributed multi-RF sensor control network for industrial automation, IEEE Wireless Telecommunications Symposium (WTS 2005), April 28–30, 2005, Pomona, CA, pp. 24–33.

[34] B.S. Prabhu, Xiaoyong Su, Harish Ramamurthy, Chi-Cheng Chu, and Rajit Gadh, WinRFID—A middleware for the enablement of Radio Frequency Identification (RFID)-based applications, in *Mobile, Wireless and Sensor Networks: Technology, Applications and Future Directions*, Rajeev Shorey, Chan Mun Choon, Ooi Wei Tsang, and A. Ananda (Eds.), John Wiley & Sons, New York, 2005.

[35] B.S. Prabhu, Xiaoyong Su, Charlie Qiu, Harish Ramamurthy, Peter Chu, and Rajit Gadh, WinRFID—Middleware for distributed RFID infrastructure, International Workshop on Radio Frequency Identification (RFID) and Wireless Sensors, Indian Institute of Technology, Kanpur, India, November 11–13, 2005.

[36] Evan Schuman, "Will Users Get Buried Under RFID Data?" Ziff Davis Internet, November 9, 2004, http://www.eweek.com/article2/0,1895,1722063,00.asp, accessed January 2007.

[37] A.D. Smith and F. Offodile, Information management of automatic data capture: An overview of technical developments, *Information Management & Computer Security*, 10, 2–3: 109–118, 2002.

[38] Anish Shah and Hiral Patel, Tag performance with Samsys 9320, RFID System Course Project, WINMEC, UCLA, Fall 2004.

[39] Seung Ryong Han and Jim Shaughhessy, Comprehensive reader and tag test (AWID), RFID System Course Project, WINMEC, UCLA, Fall 2004.

[40] Rajit Gadh and B.S. Prabhu, Radio frequency identification of Katrina hurricane victims, *IEEE Signal Processing*, March 2006.

Chapter 3

RFID Data Warehousing and Analysis

Hector Gonzalez and Jiawei Han

Contents

3.1   Introduction

Radio Frequency IDentification (RFID) is a technology that allows a sensor (RFID reader) to read, from a distance and without line of sight, a unique identifier that is provided (via a radio signal) by an "inexpensive" tag attached to an item. RFID offers a possible alternative to barcode identification systems and it facilitates applications like item tracking and inventory management in the supply chain. The technology holds the promise to streamline supply chain management, facilitate routing and distribution of products, and reduce costs by improving efficiency.

Large retailers like Wal-Mart, Target, and Albertsons have already begun implementing RFID systems in their warehouses and distribution centers, and are requiring their suppliers to tag products at the pallet and case levels. Individual tag prices have been steadily falling, and, in the near future, it is expecting to see item-level tagging for many products. The main challenge then becomes how can companies handle and interpret the enormous volume of data that a RFID application generates. Venture Development Corporation [20], a research firm, predicts that when tags are used at the item level, Wal-Mart will generate nearly 7 terabytes of data every day. Database vendors like Oracle, IBM, Teradata, and some startups are beginning to provide solutions to integrate RFID information into their enterprise data warehouses.

Example: Suppose a retailer with 3,000 stores sells 10,000 items a day per store. Assume that we record each item movement with a tuple of the form (*EPC location time*), where the electronic product code (EPC) uniquely identifies each item.* If each item leaves only 10 traces before leaving the store by going through different locations, this application will generate at least 300 million tuples per day. A manager may make queries on the duration of paths, like (Q_1): "List the average shelf life of dairy products in 2003 by manufacturer," or on the structure of the paths, like (Q_2): "What is the average time that it took coffeemakers to move from the warehouse to the shelf and finally to the checkout counter in January of 2004?" New data structures and algorithms need to be developed that will provide fast responses to such queries even in the presence of terabyte-sized data.

Such enormous amounts of low-level data and flexible high-level queries pose great challenges to traditional relational and data warehouse technologies because the processing may involve retrieval and reasoning over a large number of interrelated tuples through different stages of object movements. No matter how the objects are sorted and clustered, it is difficult to support various kinds of high-level queries in a uniform and efficient way. A nontrivial number of queries may even require a full scan of the entire RFID database.

In order to conduct efficient analysis over large RFID data sets it is important to eliminate data redundancy, perform lossless data compression by aggregating bulky object movements, and perform lossy data compression by registering data at high levels of abstraction.

3.1.1 Redundancy Elimination

It is important to eliminate the redundancy present in RFID data. Each reader provides tuples of the form (*EPC location time*) at fixed time intervals. When an item stays at the same location for a period of time, multiple tuples will be generated. These tuples can be grouped into a single one of the form (*EPC, location, time_in, time_out*). For example, if a supermarket has readers on each shelf that scan the items every minute, and items stay on the shelf on average for one day, we get a 1,440 to 1 reduction in size without loss of information.

3.1.2 Bulky Movement Compression

Items tend to move and stay together through different locations. For example, a pallet with 500 cases of compact discs (CDs) may arrive at the warehouse; from

---

* We will use the terms EPC and RFID tag interchangeably throughout the chapter.

there, cases of 50 CDs may move to the shelf; and from there, packs of 5 CDs may move to the checkout counter. A single *stay* tuple of the form (*EPC_list, location, time_in, time_out*) can be used to register the CDs that arrive in the same pallet and stay together in the warehouse, and, thus, generate a 80 percent space saving.

An alternative compression mechanism is to store a single *transition* record for the 50 CDs that move together from the warehouse to the shelf, i.e., to group transitions and not stays. The problem with transition compression is that it makes it difficult to answer queries about items at a given location or going through a series of locations. For example, in order to answer the query: *What is the average time that CDs stay at the shelf?* you can directly get the information from the *stay* records with *location = shelf,* but if only *transition* records are available, you need to find all the transition records with *origin = shelf* and the ones with *destination = shelf* join them on the EPC and compute *departure time–arrival time.* Another method of compression would be to look at the sequence of locations that an item goes through as a string, and use a Trie [6] to store the strings. The Trie data structure is a tree where each node is associated with a string prefix, and all descendant of the node share the prefix. The problem with this approach is that you lose compression power. In the CDs example, if the 50 items all stay at the warehouse together, but they come from different locations, a trie would have to create distinct nodes for each, thus gaining no compression.

3.1.3 EPC List Compression

Further compression can be gained by reducing the size of the EPC lists in the *stay* records by grouping items that move to the same location. For example, if you have a *stay* record for the 50 CDs that stayed together at the warehouse, and later these CDs moved in two groups to the shelf and truck locations, it is possible to replace the list of 50 EPCs in the stay record for just two *generalized identifiers* (*gids*), which in turn point to the concrete EPCs. In this example, we will store a total of 50 EPCs, plus two gids, instead of 100 EPCs (50 in the warehouse, 25 on shelf, 25 in truck). In addition to the compression benefits, query processing speedup is gained by assigning path-dependent names to the gids. In the CDs example, you could name the gid for the warehouse 1, and the gid for the shelf 1.1, and for the truck 1.2. If you get a query asking for the average time to go from the warehouse to the shelf for CDs, instead of intersecting the EPC lists for the *stay* records at each location, you can directly look at the gid name and determine if the EPCs are linked. The path-dependent naming scheme provides the benefits of a tree structure representation of a trie without taking a significant compression penalty.

3.1.4 Aggregation Along Concept Hierarchies

Most queries are likely to be at a high level of abstraction, and will only be interested in the low-level individual items if they are associated with some interesting patterns discovered at a high level. For example, query (Q_1) asks about dairy products by a manufacturer. It is possible after seeing the results, the user may make subsequent queries and drill down to individual items. Significant compression can be achieved by creating the *stay* records not at the raw level, but at a minimal level of abstraction shared by most applications while keeping pointers to the RFID tags. This allows the system to operate on a much smaller dataset, fetching the original data only when absolutely necessary.

3.2 RFID Data

Data generated from a RFID application can be seen as a stream of RFID readings of the form (*EPC, location, time*), where *EPC* is the unique identifier read by a RFID reader, *location* is the place where the RFID reader scanned the item, and *time* is when the reading took place. Tuples are usually stored according to a time sequence. A single EPC may have multiple readings at the same location, each reading is generated by the RFID reader scanning for tags at fixed time intervals. Table 3.1 is an example of a raw RFID database where a symbol starting with *r* represents a RFID tag, *l* a location, and *t* a time. The total number of records in this example is 188.

In order to reduce the large amount of redundancy in the raw data, data cleaning should be performed. The output after data cleaning is a set of clean *stay* records of the form (*EPC, location, time_in, time_out*) where *time_in* is the time when the object enters the location and *time_out* is the time when the object leaves the location.

Data cleaning of *stay* records can be accomplished by sorting the raw data on EPC and time, and generating *time_in* and *time_out* for each location by merging consecutive records for the same object staying at the same location. Table 3.2 presents the RFID database of Table 3.1 after cleaning. It has been reduced from 188 records to just 17.

Table 3.1   Raw RFID Records

Raw Stay Records
(r1, l1, t1)   (r2, l1, t1)   (r3, l1, t1)   (r4, l1, t1)   (r5, l1, t1)   (r6, l1, t1)
(r7, l1, t1) ... (r1, l1, t9)   (r2, l1, t9)   (r3, l1, t9)   (r4, l1, t9) ... (r1, l1, t10)
(r2, l1, t10)   (r3, l1, t10)   (r4, l1, t10)   (r7, l4, t10) ... (r7, l4, t19) ...
(r1, l3, t21)   (r2, l3, t21)   (r4, l3, t21)   (r5, l3, t21) ... (r6, l6, t35) ...
(r2, l5, t40)   (r3, l5, t40)   (r6, l6, t40) ... (r2, l5, t60)   (r3, l5, t60)

Table 3.2   A Cleansed RFID Database

EPC	Stay (EPC, location, time_in, time_out)		
r1	(*r1, l1, t1, t10*)	(*r1, l3, t20, t30*)	
r2	(*r2, l1, t1, t10*)	(*r2, l3, t20, t30*)	(*r2, l5, t40, t60*)
r3	(*r3, l1, t1, t10*)	(*r3, l3, t20, t30*)	(*r3, l5, t40, t60*)
r4	(*r4, 1, t1, t10*)		
r5	(*r5, l2, t1, t8*)	(*r5, l3, t20, t30*)	(*r5, l5, t40, t60*)
r6	(*r6, l2, t1, t8*)	(*r6, l3, t20, t30*)	(*r6, l6, t35, t50*)
r7	(*r7, l2, t1, t8*)	(*r7, l4, t10, t20*)	

3.3 Architecture of the RFID Warehouse

Suppose that the cleansed RFID data is viewed as a fact table with dimensions (*EPC, location, time_in, time_out:measure*). A traditional data cube will compute all possible groupings (group-bys) on this fact table by aggregating records that share the same values (or any *) at all possible combinations of dimension. If you use count as *measure*, you can get, for example, the number of items that stayed at a given location for a given month. The problem with this form of aggregation is that it does not consider links between the records. For example, if you want to get the number of items of "dairy product" that traveled from the distribution center in Chicago to stores in Urbana, the information is lost. You have the count of "dairy products" for each location, but you do not know how many of these items went from the first location to the second. A more powerful cubing model capable of aggregating data while preserving its path-like structure is needed.

A new model of RFID warehouse architecture should contain a fact table, *stay*, composed of cleansed RFID records; an information table, *info*, that stores path-independent information for each item, i.e., SKU information that is constant regardless of the location of the item such as manufacturer, lot number, color, etc.; and a *map* table that links together different records in the fact table that form a path. Figure 3.1 shows a logical view into the RFID warehouse schema. The *stay*, *info*, and *map* tables aggregated at a given abstraction level are called a *RFID-Cuboid*.

The main difference between the RFID warehouse and a traditional warehouse is the presence of the map table linking records from the fact table (*stay*) in order to preserve the original structure of the data.

The computation of RFID-Cuboids is more complex than that of regular cuboids as data aggregation has to preserve the structure of the paths traversed by EPCs at different abstraction levels. From the data storage and query processing point of view, the RFID warehouse can be viewed as a multilevel database, organized as in Figure 3.2. The raw RFID repository resides at the lowest level;

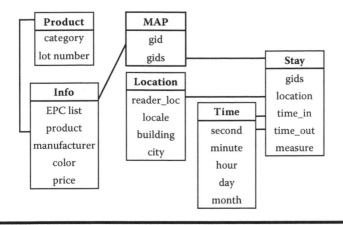

Figure 3.1   RFID Warehouse—Logical Schema.

on its top are the cleansed RFID database, the minimum abstraction level RFID-Cuboids, and a sparse subset of the full cuboid lattice composed of frequently queried (popular) RFID-Cuboids.

3.3.1   Key Ideas of RFID Data Compression

Even with the removal of data redundancy from RFID raw data, the cleansed RFID database usually is still enormous. In this section, we will elaborate on the concepts

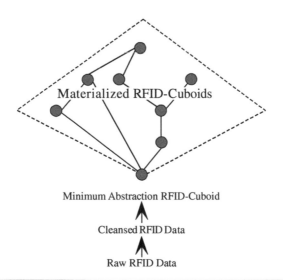

Figure 3.2   The architecture of a RFID warehouse.

of compression presented in the introduction, as these are essential to understand the RFID-Cuboid structure.

3.3.1.1   Taking Advantage of Bulky Object Movements

Since a large number of items travel and stay together through several stages, it is important to represent such a collective movement by a single record no matter how many items were originally collected. As an example, if 1,000 boxes of milk stayed in location loc_A between time t_1 (*time_in*) and t_2 (*time_out*), it would be advantageous if only one record is registered in the database rather than 1,000 individual RFID records. The record would have the form: (*gid, prod, loc_A*, t_1, t_2, 1000), where 1,000 is the count, *prod* is the product id, and *gid* is a generalized id, which will not point to the 1,000 original EPCs, but instead point to the set of new gids that the current set of objects move to. For example, if this current set of objects were split into 10 partitions, each moving to one distinct location, *gid* will point to 10 distinct new gids, each representing a record. The process iterates until the end of the object movement where the concrete EPCs will be registered. By doing so, no information is lost, but the number of records to store such information is substantially reduced.

3.3.1.2   Taking Advantage of Data Generalization

Since many users are only interested in data at a relatively high abstraction level, data compression can be explored to group, merge, and compress data records. For example, if the minimal granularity of time is hours, then objects moving within the same hour can be seen as moving together and be merged into one movement. Similarly, if the granularity of the location is shelf, objects moving to the different layers of a shelf can be seen as moving to the same shelf and be merged into one. Similar generalizations can be performed for products (e.g., merging different sized milk packages) and other data as well.

3.3.1.3   Taking Advantage of the Merge or Collapse of Path Segments

In many analysis tasks, certain path segments can be ignored or merged for simplicity of analysis. For example, some nonessential object movements (e.g., from one shelf to another in a store) can be completely ignored in certain data analysis. Some path segments can be merged without affecting the analysis results. For store managers, merging all the movements before the object reaches the store could be desirable. Such merging and collapsing of path segments may substantially reduce the total size of the data and speed up the analysis process. Figure 3.3 presents two different path collapsing schemes. The path in the top is the view of a store manager

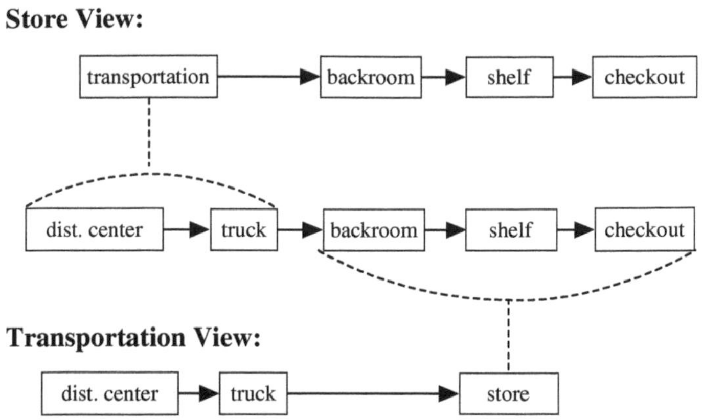

Store View:

Transportation View:

Figure 3.3   Path collapsing.

that is only concerned with detailed movements inside the store. The path at the bottom is the view of a transportation manager, only interested in detailed movements related to transportation locations.

3.3.2 RFID-CUBOID

With the data compression principles in mind, the RFID-Cuboid can be defined. It is a data structure for storing aggregated data in the RFID warehouse. Its design ensures that the data is disk-resident, summarizing the contents of a cleansed RFID database in a compact yet complete manner while allowing efficient execution of both online analytical processing (OLAP) and tag-specific queries.

The RFID-Cuboid consists of three tables: (1) *Info*, which stores product information for each RFID tag; (2) *Stay*, which stores information on items that stay together at a location; and (3) *Map*, which stores path information necessary to link multiple *stay* records.

3.3.2.1 Information Table

In addition to the data provided by the RFID readers, it is important to store additional information related to each item, such as product, manufacturer, description, and production batch. This information can be an important part of the warehousing application as users may want to aggregate data on the paths traveled by items from a particular manufacturer and produced at a given manufacturing plant (this may be useful in determining the source of a defective product). The data contained in the information table is path independent, i.e., it remains constant regardless of the location of the item. Each dimension of the information table can itself have an

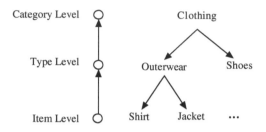

Figure 3.4 Product hierarchy.

associated concept hierarchy. Figure 3.4 presents an example concept hierarchy for the product dimension.

The information table (*info*) contains a set of attributes that provide a path-independent extended description of RFID tags. Each entry in *Info* is a record of the form: ⟨ (*EPC_list*), $(d_1,...,d_m)$: $(m_1,..., m_i)$ ⟩, where the code list contains a set of items (i.e., rfids) that share the same values for dimensions $d_1,...,d_m$, and $m_1,..., m_i$ are measures of the given items (e.g., price).

3.3.2.2 Stay Table

As mentioned in the introduction, items tend to move and stay together through different locations. Compressing multiple items that stay together at a single location is vital in order to reduce the enormous size of the cleansed RFID database. In real applications, items tend to move in large groups. At a distribution center there may be tens of pallets staying together, and then they are broken into individual pallets at the warehouse level. Even if products finally move at the individual item level from a shelf to the checkout counter, the *stay* compression will save space for all previous steps taken by the item.

The *stay* table (*stay*) contains an entry for each group of items that stay together at a certain location. Each entry in *stay* is a record of the form: ⟨(*gids, location, time_in, time_out*): $(m_1,..., m_k)$⟩, where *gids* is a set of generalized record ids, each pointing to a list of RFID tags or lower level *gids;location* is the location where the items stayed together; *time_in* is the time when the items entered the location; and *time_out*, the time when they leave. If the items did not leave the location, *time_out* is *NULL*. $m_1,..., m_n$ are the measures recorded for the stay (e.g., count, average time at *location*, and the maximal time at *location*).

Stay records may contain items that belong to different categories of products, for example, if milk and cheese stay together at the warehouse and share the same measure, they will have only one stay record for the warehouse location. Items that stay together, share the same path, and belong to the same info record have their measures aggregated into a single *stay* record, for example, if you have ten boxes of milk and two cans of cheese traveling together through the supply chain, and the information table is aggregated at the product category level, in this case dairy, you can aggregate the measure of the ten boxes of milk and two cans of cheese at each stage and record a single *stay* record. But if the information table is aggregated at the product level and the aggregate measure of the milk and cheese were different, you would need two *stay* records per stage.

3.3.2.3   Map Table

The *map* table is an efficient structure that allows query processing to link together stages that belong to the same path in order to perform structure-aware analysis, which could not be answered by a traditional data warehouse. There are two main reasons for using a *map* table instead of recording the complete EPC lists at each stage: data compression and query processing efficiency.

First, it is very expensive to record each RFID tag on the EPC list for every *stay* record it participated in. For example, if we assume that 10,000 items move in the system in groups of 10,000, 1,000, 100, and 10 through four stages, instead of using 40,000 units of storage for the EPCs in the *stay* records, the *map* table uses only 1,111 units* (1,000 for the last stage, 100, 10, and 1 for the ones before).

The second and the most important reason for having such a *map* table is the efficiency in query processing. Suppose each map entry were given a path-dependent label. To compute, for example, the average duration for milk to move from the distribution center (*D*), to the store backroom (*B*), and finally to the shelf (*S*), you need to locate the *stay* records for milk at each stage. To get three sets of records *D*, *B*, and *S*, one has to intersect the EPC lists of the records in *D* with those in *B* and *S* to get the paths. By using the *map*, the EPC lists can be orders of magnitude shorter and, thus, reduce input/output (I/O) costs. Additionally, if path-dependent naming of the *map* entries is used, you can compute list intersections much faster.

The *map* table contains mappings from higher level *gids* to lower level ones or EPCs. Each entry in *map* is a record of the form: $\langle gid, (gid_1, ..., gid_n) \rangle$, meaning that *gid* is composed of all the EPCs pointed to by $gid_1, ..., gid_n$. The lowest level *gids* will point directly to individual items.

In order to facilitate query processing, the naming scheme assigns path-dependent labels to high level *gids*. The label contains one identifier per location traveled by the items in the *gid*. Figure 3.5 presents the gid map for the cleansed data in Table 3.2. It can be seen that the group of items with *gid* 0.0 arrive at *l*1, and then subdivide into 0.0.0, and s0.0; the items in s0.0 stay in location *l*1 while the items in 0.0.0 travel to *l*3, and subdivide further into 0.0.0.0 and s0.0.0; the items in s0.0.0 stay at location *l*3, while the ones in 0.0.0.0 move to location *l*5.

---

* This figure does not include the size of the map itself, which should use 12,221 units of storage, still much smaller than the full EPC lists.

Map – Tabular View

0	0.0,0.1
0.0	0.0,0.0,s0.0
0.0.0	0.0.0.0,s0.0.0
0.0.0.0	r2,r3
s0.0	r4
s0.0.0	r1
0.1	0.1.0,0.1.1
0.1.0	0.1.0.0,0.1.0.1
0.1.0.0	r5
0.1.0.1	r6
0.1.1	r7

Map – Graphical View

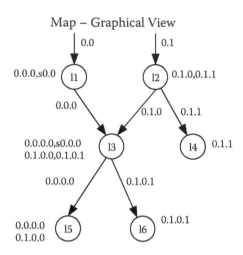

Figure 3.5   GID map.

3.3.3 Lattice of RFID-Cuboids

Each dimension in the *stay* and *info* tables has an associated concept hierarchy, which is a partial order of mappings from lower levels of abstraction to higher ones. The lowest corresponds to the values in the raw RFID data stream itself, and the associated per item information, which could be at the stock keeping unit (SKU) level. The highest is *, which represents any value of the dimension.

In order to provide fast response to queries specified at various levels of abstraction, it is important to precompute some RFID-Cuboids at different levels of the concept hierarchies for the dimensions of the *info* and *stay* tables. It is obviously too expensive to compute all the possible generalizations, so partial materialization is a preferred choice. This problem is analogous to determining which set of cuboids in a data cube to materialize in order to answer OLAP queries efficiently given the limitations on storage space and precomputation time. This issue has been studied extensively in data cube research [2][13], and the principles are generally applicable to the selective materialization of RFID-Cuboids.

RFID-Cuboids are materialized at the minimal interesting level at which users will be interested in querying the database, and a small set of higher level structures that are frequently requested and that can be used to quickly compute nonmaterialized RFID-Cuboids.

An RFID-Cuboid residing at the minimal interesting level are computed directly from the cleansed RFID database and is the lowest (most detailed) cuboid that can be queried unless one digs directly into the cleansed data in some very special cases.

3.4 Construction of an RFID Warehouse

In order to construct the RFID-Cuboid, as described in the chapter, one needs a compact data structure that does the following efficiently: (1) assign path-dependent labels to the gids, (2) minimize the number of output *stay* records while computing aggregates that preserve the path-like nature of the data, and (3) identify path segments that can be collapsed. Such structure is a tree representation of the path database, where each node in the tree represents a path stage; and all common path prefixes in the database share the same branch in the tree. The tree is traversed in breath-first order, that is, we visit nodes level by level to assign path-dependent labels to the nodes. The minimum number of output *stay* records can be quickly determined by aggregating the measures of all the items that share the same branch in the tree. And path segments can be collapsed by simply merging parent/child nodes that correspond to the same location, or by merging uninteresting locations. Additionally, the tree can be constructed by doing a single scan of the cleansed RFID database, and it can be discarded after one has materialized the output *info*, *stay*, and *map* tables.

The most common implementation of the group-by operator, which would sort the input records on *location*, *time_in*, and *time_out*, and generate a list of RFID tags that share the same values on these dimensions, would fail in generating the RFID-Cuboid in an efficient manner. The problem with this approach is that one loses the ability to determine which of the items that share the same *location*, *time_in*, and *time_out* values can actually have their measures aggregated into a single value. The reason is that one cannot quickly determine which subset of all the items followed the same path throughout the system without doing an unmanageable number of item list intersections. For this same reason, one cannot easily construct the *map* table, assign path-dependent labels to the gids, or determine if two different records can be collapsed into one.

Algorithm 1 summarizes the method for constructing an RFID-Cuboid from the cleansed RFID database. It takes as input the clean *stay* records *S*, the *info* records *I* describing each item, and the level of abstraction for each dimension *L*. The output of the algorithm is an RFID-Cuboid represented by the *stay*, *info*, and *map* tables aggregated at the desired abstraction level.

First, the information table is aggregated to the level of abstraction specified by *L*. This is done by using a regular cubing algorithm. Second, the BuildPathTree procedure is called (Algorithm 2), which constructs a tree for the paths traveled by items in the cleansed database. The paths that have the same branch in the tree share the same path prefix. Each node in the tree is of the form (*location*, *time_in*, *time_out*, *measure_list*, *children_list*) where *time_in* is the time at which the items entered *location*, *time_out* the time at which they left, *measure_list* contains the *stay* measures (from the parent's location to this node's location) for every item that stayed at the node, and *children_list* is a list of nodes where the items in the node moved next. All processing from this point on in the algorithm will be done using the tree, which is called *path_tree*.

Third, the algorithm merges consecutive nodes in the tree that correspond to the same location. This is done because it is possible for two distinct locations to be aggregated to the same higher level location. This step achieves compression by collapsing paths to the given abstraction level. Fourth, gids are generated for the nodes in the tree. This is done in the breath-first order, and each node receives a unique id that is appended to the gid of its parent node. This naming scheme is used to speed up the computation of linked *stay* records for query processing. Fifth, the *measure_list* at each node is compacted to one aggregate measure for each group of RFID tags that share the same leaf (descendant from the node) and *info* record. Further compression is done when several leaves share the same measure.

Finally, the algorithm traverses the tree, generating the new *stay* table: Each node generates as many records as the number of distinct measures it contains. It is possible for multiple nodes to share the same *stay* record if they share all the attributes and measures. For example, if there are two nodes in the path_tree that have the same location, *time_in*, *time_out*, and measures, we generate a single *stay* record that has two gids, one for each node.

Table 3.3 presents the *stay* table that would be generated by running Algorithm 1 on the cleansed data from Table 3.2. We have now gone from 188 records in the raw data, to 17 in the cleansed data, and then to 7 in the compressed data.

3.4.1 Construction of Higher Level RFID-CUBOIDs from Lower Level Ones

Once the minimum abstraction level RFID-Cuboid has been constructed, it is possible to gain efficiency by constructing higher level RFID-Cuboids starting from the existing RFID-Cuboid instead of directly from the cleansed RFID data. This can be accomplished by running Algorithm 1 with the *stay* and *info* tables of the lower level RFID-Cuboid as input, and using the *map* table of the input RFID-Cuboid to expand each gid to the EPCs it points to. The benefit of such an approach is that the input *stay* and *info* tables can be significantly smaller than the cleansed tables, thus, greatly gaining in space and time efficiency.

Table 3.3   Output Stay Records

gid	loc	t1	t2	count	measure
0.0	l1	t1	t10	4	9
0.1	l2	t1	t8	3	7
0.0.0	l3	t20	t30	3	9
0.1.0	l3	t20	t30	2	19
0.1.1	l4	t10	t20	1	19
0.0.0.0,0.1.0.0	l5	t40	t60	3	19
0.1.0.1	l6	t35	t50	1	14

Table 3.4   Data Set Parameters

$\beta = (s_1,...,s_k)$	Path Bulkiness
k	Average path length
P	Number of products
N	Number of cleansed RFID records

Algorithm 1: BuildCuboid

5in

Input: *Stay* records S, Information records I, aggregation level L
Output: RFID-Cuboid
Method:

1: I' = aggregate I to L
2: path_tree = BuildPathTree (S, L)
3: Merge consecutive nodes with the same location in the path_tree
4: Traverse path_tree in the breath-first order and assign gid to each node (gid = parent.gid + '.' + unique id) where the parent/children gid relation in the path_tree defines the output MAP
5: Compute aggregate measures for each node, one per group of items with the same information record and in the same leaf
6: Create S' by traversing path_tree in the breath-first order and generating *stay* records for each node (multiple nodes can contribute to the same stay record)
7: Output MAP, S', and I'

Observation: Given a cleansed *stay* and *info* input tables, the RFID-Cuboid structure has the following properties:

1. **Construction cost:** The RFID-Cuboid can be constructed by doing a single sequential scan on the cleansed stay table.
2. **Completeness:** The RFID-Cuboid contains sufficient information to reconstruct the original RFID database aggregated at abstraction level L.
3. **Compactness:** The number of records in the output *stay* and *info* tables is no larger than the number of records in the input *stay* and *info* tables, respectively.

Rationale: The first property has been shown in the RFID-Cuboid construction algorithm. The completeness property can be proved by using the *map* table to expand the gids for each *stay* record to get the original data. The compactness property is proved noticing that Algorithm 1 emits, at most, one output record per distinct measure per node, and the number of distinct measures in a node is limited by the number of input *stay* records. The size of the output *info* table, by definition of the group-by operation, is bound by the size of the input *info* table.

Algorithm 2: BuildPathTree

Input: Stay *S*, aggregation level *L*
Output: path_tree
Method:

   1: root = new node
   2: **for** each record *s* in *S* **do**
   3:    *s*'= aggregate s to level L
   4:    parent = lookup node for *s*'.rfid
   5:    **if** parent == NULL **then**
   6:       parent = root
   7:    **end if**
   8:    node = lookup *s*'.rfid in parent's children
   9:    **if** node == NULL **then**
10:       node = new node
11:       node.loc = s.loc
12:       node.t1 = s.t1
13:       node.t2 = s.t2
14:       add node to parent's children
15:    **end if**
16:    node.measure_list += ⟨*s*'.gid,*s*'.measure⟩
17: **end for**
18: return root

3.5 Query Processing

A very important aspect of the RFID warehouse is the efficient handling of traditional OLAP operations, i.e., drill-down, roll-up, slice, and dice. These operations allow a data analyst to navigate the lattice of *RFID-Cuboids*. The warehouse also needs to support a new type of operation, path selection, which is unique to RFID data sets and is necessary to compute path dependent aggregates.

3.5.1 OLAP

Traditionally a data cube is navigated by the usage of four operations:

3.5.1.1 Rollup

The rollup operation performs aggregation on a data cube, either by climbing up a concept hierarchy for a dimension or by dimension reduction. For example, assume that the location has the concept hierarchy *location* → *locale* → *location category*, time has the hierarchy *second* → *hour* → *day*, and product has the hierarchy *EPC* →

SKU. Now, assume that one is looking at the *stay* table at the level of ⟨*locale, hour*⟩ and at the *info* table at the level of *EPC*, i.e., for each *EPC*, one aggregates *stay* with the same *time_in*, and *time_out* at the hour level, and the same location at the *locale* level. One can rollup to a higher abstraction level, such as, ⟨*location, category, day*⟩ and *SKU* where each *stay* and *info* perform coarser aggregation.

The rollup operation is implemented in the RFID warehouse by retrieving the relevant RFID-Cuboid, or if the cuboid has not been materialized by computing it on the fly from the closest materialized RFID-Cuboid that is at a lower abstraction level than the needed one.

3.5.1.2   Drill Down

This operation is the inverse of rollup, during drill down one moves to more detailed data by climbing down a concept hierarchy or by adding new dimensions. Using the above example, we may drill down from ⟨*locale, day,*⟩, *SKU* to ⟨*locale, second*⟩, *EPC*.

The drill-down operation is implemented in the RFID warehouse in the same manner as the rollup one. We just retrieve the relevant RFID-Cuboid or materialize it on the fly from the available ones. One important aspect that needs to be mentioned, is that when the minimum abstraction level RFID-Cuboid has been reached through drill-down operations, if users want to go to even more detailed data, they can retrieve the paths for the EPCs for which more detail is needed. For example, even though the lowest level RFID-Cuboid is aggregated at the SKU level, a store manager may want to retrieve the concrete EPCs for those perishable items that have stayed for more than three weeks in the refrigerator.

3.5.1.3   Slice and Dice

These operations perform selection on one or more dimensions of the RFID-Cuboid. For example, after retrieving a certain RFID-Cuboid, one may want to look only at those *stay* records that correspond to electronic goods, for the Northeast region.

The slice and dice operations can be implemented quite efficiently by using relational query execution and optimization techniques. An example of the dice operation could be: "Give me the average time that milk stays at the shelf in store S_1 in Illinois." This can be answered by the relational expression:

$$\sigma_{stay.location = shelf', info.product = milk'} \, (stay \bowtie_{gid} info).$$

3.5.2   Path Query Processing

Path queries, which ask about information related to the structure of object traversal paths, are unique to the RFID warehouse because the concept of object movements is not modeled in traditional data warehouses. It is essential to allow users to inquire about an aggregate measure computed based on a predefined sequence of locations (path). One such example could be: "What is the average time for milk to travel from farms to stores in Illinois?"

Queries on the paths traveled by items are fundamental to many RFID applications and will be the building block on top of which more complex data-mining operators can be implemented. We will illustrate this point with two real examples. First, the United States government is currently in the process of requiring the containers arriving into the country by ship to carry an RFID tag. The information can be used to determine if the path traveled by a given container has deviated from its historic path. This application may need to first execute a path-selection query across different time periods, and then use outlier detection and clustering to analyze the relevant paths. Second, plane manufacturers are planning to tag airplane parts to better record maintenance operations on each part. Again, using path selection queries one could easily associate a commonly defective part with a path that includes a certain set of suppliers who provide raw material for its construction and are likely the sources of the defect. This task may need frequent item set counting and association mining on the relevant paths. These examples show that path selection can be crucial to successful RFID data analysis and mining, and it is important to design good data structures and algorithms for efficient implementation.

More formally, a path selection query is of the form:

$$q \leftarrow \langle \sigma_c info, (\sigma_{c_1} stage_i, \ldots, \sigma_{c_k} stage_k) \rangle,$$

where $\sigma_c info$ means the selection on the *info* table based on condition c, and $\sigma_{c_i} stage_i$ means the selection based on condition c_i on the *stay* table $stage_i$. The result of a path selection query is a set of paths whose stages match the stage conditions in the correct order (possibly with gaps), and whose items match the condition c. The query expression for the example path query presented above is $c \leftarrow \langle product = \text{"milk"} \rangle$, $c_1 \leftarrow \langle location = \text{"farm"} \rangle$, and $c_2 \leftarrow \langle location = \text{"store"} \rangle$. We can compute aggregate measures on the results of a path selection query, e.g., for the example query, the aggregate measure would be average time.

Algorithm 3 illustrates the process of selecting the gids matching a given query. It first selects the gids for the *stay* records that match the conditions for the initial and final stages of the query expression. For example, g_{start} may look like $\langle 1.2, 8.3.1, 3.4 \rangle$, and g_{end} may look like $\langle 1.2.4.3, 4.3, 3.4.3 \rangle$. Then it computes the pairs of gids from g_{start} that are a prefix of a gid in g_{end}. Continuing with the example, one gets the pairs $\langle (1.2, 1.2.4.3), (3.4, 3.4.3) \rangle$. For each pair the algorithm then retrieves all the *stay* records. The pair (1.2, 1.2.4.3) would require the retrieval of all *stay* records that include gids 1.2, 1.2.4, and 1.2.4.3. Finally, a verification phase is conducted where we check that each retrieved record matches the selection conditions for each $stage_i$ and for *info*, and add those paths to the answer set.

If statistics on query selectivity are available, it may be possible to find a better optimized query execution plan than that presented in Algorithm 3. If one has a sequence of stages (*stage$_i$*, ..., *stage$_k$*), one could retrieve the records for the most selective stages, in addition to retrieving the *stay* records for *stage$_i$* and *stage$_k$*, in order to further prune the search space.

Algorithm 3: PathSelection

Input: $q \leftarrow \langle \sigma_c info, (\sigma_{c_1} stage_i, ..., \sigma_{c_k} stage_k) \rangle$, RFID warehouse
Output: the paths that match query conditions, q
Method:
1: g_{start} = select gids of *stay* records matching the condition at *stage$_1$*
2: g_{end} = select gids of *stay* records matching the condition at *stage$_k$* and that for *info*
3: **for** every pair of gids (s, e) in g_{start}, g_{end} such that s is a prefix of e **do**
4:     *path* = retrieve *stay* records for all gids from s to e
5:     **if** the stay records in *path* match conditions for *info* and for the remaining stages **then**
6:         *answer* = *answer* + *path*
7:     **end if**
8: **end for**
9: return *answer*

Analysis: Algorithm 3 can be executed efficiently if the following indices are available. A one-dimensional index for each dimension of the *stay* table, a one-dimensional index on the gid dimension of the *map* table, and a one-dimensional index on the EPC dimension of the info table. The computation of g_{start} and g_{end} can be done by retrieving the records that match each condition at *stage$_i$* and *stage$_k$*, and intersecting the results. Verification of the condition $\sigma_c info$ is done by using the *map* table to retrieve the base gids for each gid in *stage$_k$* that has as prefix a gid in *stage$_i$*. The info record for each base_gid can be retrieve efficiently by using the EPC index on the info table. When no indices are available, algorithm 3 can still be executed efficiently by performing a single scan of the *stay* and *info* tables.

3.6 Performance Study

In this section, a thorough performance analysis of the RFID warehouse is conducted. All experiments were implemented using C++ and were conducted on an Intel Xeon 2.5 GHz (512 kb L2 cache) system with 3 Gb of RAM. The system ran Red Hat Linux with the 2.4.21 kernel and gcc 3.2.3.

3.6.1 Data Synthesis

The RFID databases in the experiments were generated using a tree model for object movements. Each node in the tree represents a set of items in a location, and an edge represents a movement of objects between locations. Items at locations near the root of the tree move in larger groups, while items near the leaves move in smaller groups. The size of the groups at each level of the tree define the bulkiness, $\beta = (s_1, s_2, ..., sk)$, where s_i is the number of objects that stay and move together at level i of the tree. By making $s_i \geq s_j$ for $i > j$, we create the effect of items moving in larger groups near the factory and distribution centers, and smaller groups at the store level. The datasets for the experiments were generated by randomly constructing a set of trees with a given level of bulkiness, and generating the cleansed RFID records corresponding to the item movements indicated by the edges in the tree.

As a notational convenience, we use the following symbols to denote certain dataset parameters.

3.6.2 RFID-Cuboid Compression

The RFID-Cuboids form the basis for future query processing and analysis. As mentioned previously, the advantage of these data structures is that they aggregate and collapse many records in the cleansed RFID database. Here, we examine the effects of this compression on different data sets. We compare two different compression strategies, both use the *stay* and *info* tables, but one uses the *map* table as described in section 3.3.2.3, whereas the other uses a tag list to record the tags at each *stay* record (nomap).

Figure 3.6 illustrates the size of the cleansed RFID database (raw) compared with the *map* and *nomap* RFID-Cuboids. The datasets contain 1,000 distinct products, traveling in groups of 500, 150, 40, 8, and 1 through 5 path stages, and 500,000 to 10 million cleansed RFID records. The RFID-Cuboid is computed at the same level of abstraction of the cleansed RFID data and, thus, the compression

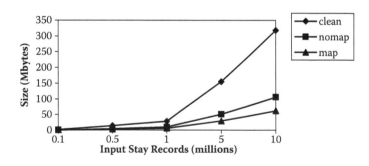

Figure 3.6   Compression versus cleansed data size: $\mathcal{P} = 1000$, $\mathcal{B} = (500, 150, 40, 8, 1)$, $\mathcal{K} = 5$.

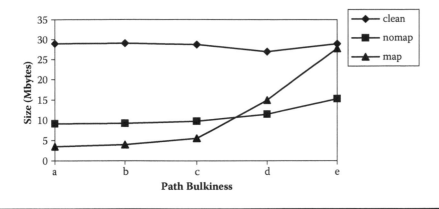

Figure 3.7 **Compression versus data bulkiness:** $\mathcal{P} = 1000$, $\mathcal{N} = 1,000,000$, $\mathcal{K} = 5$.

is lossless. As can be seen the Figure 3.6, the RFID-Cuboid that uses the *map* has a compression power of around 80 percent, while the one that uses tag lists has a compression power of around 65 percent. The benefit of the *map* comes from the fact that it avoids registering each tag at each location. In both cases, the compression provided by collapsing *stay* records is significant.

Figure 3.7 also shows the size of the cleansed RFID database (raw) compared with the *map* and *nomap* RFID-Cuboids. In this case, we vary the degree of bulkiness of the paths, e.g., the number of tags that stay and move together through the system. We define five levels of bulkiness: $a = (500, 230, 125, 63, 31)$, $b = (500, 250, 83, 27, 9)$, $c = (500, 150, 40, 8, 1)$, $d = (200, 40, 8, 1, 1)$, and $e = (100, 10, 1, 1, 1)$; the bulkiness decreases from dataset *a* to *e*. As can be seen in the figure, for more bulky data, the RFID-Cuboid that uses the *map* clearly outperforms the *nomap* cuboid; as we move towards less bulky data, the benefits of the *map* decrease as there are many entries in the *map* that point to just one gid. For paths where a significant portion of the stages are traveled by a single item, the benefit of the map disappears and one is better off using tag lists. A possible solution to this problem is to compress all *map* entries that have a single child into one.

Figure 3.8 shows the compression obtained by climbing along the concept hierarchies of the dimensions in the *stay* and *info* tables. Level-0 cuboids have the same level in the hierarchy as the cleansed RFID data. The three higher level cuboids offer one, two, and three levels of aggregation, respectively, at all dimensions (location, time, product, manufacturer, color). As expected, the size of the cuboids at higher levels decreases. In general, the cuboid using the *map* is smaller, but for the top most level of abstraction, the size is the same as for the nomap cuboid. At level 3 the size of the *stay* table is just 96 records, and most of the space is actually used by recording the RFID tags themselves and, thus, it makes little difference if one uses a *map* or not.

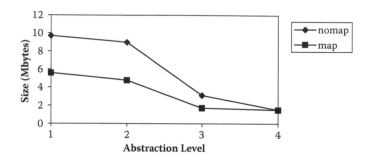

Figure 3.8 Compression versus abstraction level: $\mathcal{P} = 1000$, $\mathcal{B} = (500, 150, 40, 8, 1)$, $\mathcal{K} = 5$, $\mathcal{N} = 1,000,000$.

Figure 3.9 shows the time to build the RFID-Cuboids at the same four levels of abstraction used in Figure 3.8. In all cases, the cuboid was constructed starting from the cleansed database. We can see that cuboid construction time does not significantly increase with the level of abstraction. This is expected as the only portion of the algorithm that incurs extra cost for higher levels of abstraction is the aggregation of the *info* table, and in this case it contains only 1,000 entries. This is common as usually the cleansed RFID *stay* table is orders of magnitude larger than the *info* table. The computation of RFID-Cuboids can also be done from lower-level cuboids instead of doing it from the cleansed database. For the cuboids 1 to 3 of Figure 3.9, one can obtain savings of 50 to 80 percent in computation time if you build cuboid *i* from cuboid *i* − 1.

3.6.3 Query Processing

A major contribution of the RFID data warehouse model is the ability to efficiently answer many types of queries at various levels of aggregation. In this section, we show this efficiency in several settings. Three query execution scenarios are

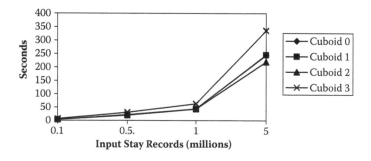

Figure 3.9 Construction time: $\mathcal{P} = 1000$, $\mathcal{B} = (500, 150, 40, 8, 1)$, $\mathcal{K} = 5$.

compared; the first is a system that directly uses the cleansed RFID database (raw), the second one uses the *stay* table but no *map*, it instead uses tag lists at each stay record (nomap); and the third is the RFID-Cuboid described above using *stay* and *map* tables (map). Assume that for each of the scenarios you have a B+tree on each of the dimensions. In the case of the *map* cuboid, the index points to a list of gids matching the index entry. In the case of the *nomap* cuboid and the cleansed database, the index points to the tuple (*RFID tag, record id*). This is necessary as each RFID tag can be present in multiple records. The query answering strategy used for the *map* cuboid is the one presented in Algorithm 3. The strategy for the other two cases is to retrieve the (*RFID tag, record id*) pairs matching each component of the query, intersecting them, and finally retrieving the relevant records.

For the experiments, assume that you have a page size of 4096 bytes and that RFID tags, record ids, and gids use 4 bytes each. We also assume that all the indices fit in memory except for the last level. For each of the experiments, you generated 100 random path queries. The query specifies a product, a varying number of locations (three on average), and a time range to enter the last stage (*time_out*). Semantically, this is equivalent to asking, "What is the average time for product X to go through locations L_i, ..., L_k entering location L_k between times $t_1 - t_2$?"

Figure 3.10 shows the effect of different cleansed database sizes on query processing. The *map* cuboid outperforms the cleansed database by several orders of magnitude and, most importantly, query answer time is independent of database size. The *nomap* cuboid is significantly faster than the cleansed data, but it suffers from having to retrieve very long RFID lists for each stage. The *map* cuboid benefits from using very short gid lists, and using the path-dependent gid-naming scheme that facilitates determining if two *stay* records form a path without retrieving all intermediate stages.

Figure 3.11 shows the effects of path bulkiness on query processing. For this experiment, we set the number of *stay* records constant at one million. The bulkiness levels are the same as those used for the experiment in Figure 3.7. As with the compression experiment, since there are more bulky paths, the *map* cuboid is an

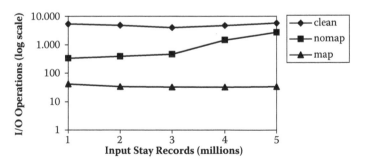

Figure 3.10   I/O cost versus cleansed data size: $\mathcal{P} = 1000$, $\mathcal{B} = (500, 150, 40, 8, 1)$, $\mathcal{K} = 5$.

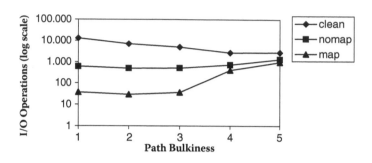

Figure 3.11 I/O cost versus path bulkiness: $\mathcal{P} = 1000$, $\mathcal{K} = 5$.

order of magnitude faster than the cleansed RFID database. As we get less bulky paths, the benefits of compressing multiple *stay* records decreases until the point at which it is no better than using the cleansed database. The difference between the *map* and *nomap* cuboids is almost an order of magnitude for bulky paths, but as in the previous case for less bulky paths the advantage of using the *map* decreases.

Figure 3.12 shows the effect of having a varying number of distinct products, 10 to 1,000. A smaller number of products corresponds to a smaller *info* table. In this case, we see that the size of the info table does not affect performance as we have a B+Tree on each of its dimensions and the size of the record lists is usually small, at most 1,000 in this case. As we said before, this is realistic since we expect to have *stay* tables that are orders of magnitude larger than *info* tables.

3.7 Discussion

In this section, we discuss the related work and explore the possible extensions of the RFID warehousing framework.

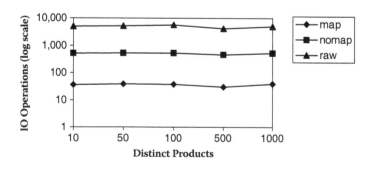

Figure 3.12 I/O cost versus distinct products: $\mathcal{B} = (500, 150, 40, 8, 1)$, $\mathcal{K} = 5$.

3.7.1 Related Work

RFID technology has been extensively studied from mostly two distinct areas: the electronics and radio communication technologies required to construct readers and tags [4], and the software architecture required to securely collect and manage online information related to tags [17][18].

More recently a third line of research dealing with mining of RFID data has emerged. Chawathe et al. [3] introduced the idea of creating a warehouse for RFID data, but do not go into the data structure or algorithmic details. Gonzalez et al. (2006c) [10] present a concrete RFID warehouse design and most of the contents of this chapter are based on their design. Gonzalez et al. (2006a) [8] present the idea of the FlowCube, which is a warehouse similar to the one introduced in [10], but where each cell stores a FlowGraph [9], which is a probabilistic workflow model that captures the major movement trends and significant deviations for items aggregated in the cell.

Cleaning of RFID data sets is another problem related to warehouse construction, as one relies on clean data to build a high-quality model. There have been extensive studies in the design of communication protocols, at the hardware level, which are robust to interference and collisions [7][14]. At the software level, the use of smoothing windows has been proposed as a method to reduce false negatives. Floerkemeier and Lampe [5] proposed the fixed-window smoothing. Jeffrey et al. (2006a) [15] proposed a variable-window smoothing that adapts window size dynamically according to tag detection rates, by using a binomial model of tag readings. Jeffrey et al. (2006b) [16] present a framework to clean RFID data streams by applying a series of filters, but do not go into the details of filter design. Gonzalez et al. (2007) [11] define a cost-conscious cleaning framework that is designed to handle the efficient and accurate cleaning of large RFID datasets; the idea of the framework is to construct a cleaning plan that applies inexpensive cleaning methods whenever possible, and more expensive cleaning techniques only when absolutely necessary.

An RFID data warehouse shares many common principles with the traditional data cube [1][2][12]. They both aggregate data at different levels of abstraction in multidimensional space. Since each dimension has an associated concept hierarchy, both can be (at least partially) modeled by a *Star* schema. The problem of deciding which RFID-Cuboids to construct in order to provide efficient answers to a variety of queries specified at different abstraction levels is analogous to the problem of partial data cube materialization studied in [13][19]. However, RFID-Cuboid differs from a traditional data cube in that it also models *object transitions* in multidimensional space, which is crucial to answering queries related to object movements/transitions as demonstrated in this analysis.

3.7.2 Open Issues

Below is a discussion of methods for incremental update in the RFID warehouse, construction of fading RFID model, and its linkage with mining RFID data.

3.7.2.1 Incremental Update

Incremental update is crucial to RFID applications. A warehouse will receive constant updates as new tags enter the system and objects move to new locations. Algorithm 1 can be applied only to the new data to generate new *stay* and *map* tables (users need to initialize the gid naming as to not generate duplicates with the existing *map*). The updated cuboid will be the union of the old *stay* and *map* tables with the new ones. The main reason that we can just run the algorithm in the new data without worrying about the old data is that the updates are only for item movements with higher timestamps than we already have, so they will form new nodes in the path tree.

3.7.2.2 Construction of a Fading RFID Warehouse Model

In most situations, the more remote in time or distance the data is, the less interest a user would have to study it in detail. For example, one may not be interested in the detailed tag movements among shelves if the data is years old, or thousands of miles away, or not in the same sector (e.g., manufacturer versus store manager). This will relieve the burden of storing the enormous size of historical or remote or unrelated RFID databases.

The RFID warehouse can be easily adapted to a *fading model* where remote historical or distantly located data can be stored with low resolution (i.e., not retaining such data at a low abstraction level). This can be done incrementally by further summarizing data being faded at an abstraction level higher than the recent and close-by data or by simply tossing some low-level cuboid and raising the level of the corresponding minimum abstraction level RFID-Cuboid for such data.

For example, one could use the following fading strategy: Summarize data for the last 30 days at the *hour* level, for the last 12 months at the *day* level, and for even older data at the *week* level. The architecture of the RFID warehouse is very well suited for such summarization schemes.

3.7.2.3 Data Mining in the RFID Warehouse

The RFID warehouse model facilitates efficient data mining since the data in the model is well structured, aggregated, and organized. Take the frequent itemset mining as an example. One can easily mine frequent itemsets at high abstraction levels since such data is already aggregated with count and other measures computed. The Apriori pruning can be used to prune the search along the infrequent high-level itemsets. Efficient methods, such as progressive deepening, can be further explored to reduce the search space. For example, if dairy products and meat are not sold together frequently, there is no need to drill down to see whether milk and beef will be sold together frequently. The RFID warehouse model naturally facilitates levelwise pruning based on its multilevel structure and levelwise precomputation.

3.7.2.4   Warehousing Scattered Movement Data

Notice that the data model presented in this chapter and the methods for warehouse construction and query analysis are based on the assumption that RFID data tends to move together in bulky mode, especially at the early stage. This fits a good number of RFID applications, such as supply chain management. However, there are also other applications where RFID data may not have such characteristics. For example, tracking of cars on a road network or people in a building may exhibit individual movements that cannot be compressed through the use of generalized identifiers. We believe that further research is needed to construct efficient models for such applications.

3.8   Conclusions

This chapter has introduced a novel model for warehousing RFID data. The design allows high-level analysis to be performed efficiently and flexibly in multidimensional space. The model is composed of a hierarchy of highly compact summaries (RFID-Cuboids) of the RFID data aggregated at different abstraction levels where data analysis can take place. Each cuboid records tag movements in the *stay*, *info*, and *map* tables that take advantage of the fact that individual tags tend to move and stay together (especially at higher abstraction levels) to collapse multiple movements into a single record without loss of information. The performance study shows that the size of RFID-Cuboids at interesting abstraction levels can be orders of magnitude smaller than the original RFID databases and can be constructed efficiently. Moreover, the empirical evaluation highlights the power of the data structures and algorithms in efficient answering of a wide range of RFID queries, especially those related to object transitions.

The model in this chapter focuses on efficient data warehousing and OLAP-styled analysis of RFID data. Efficient methods for a multitude of other data mining problems for the RFID data (e.g., trend analysis, outlier detection, path clustering) remain open and should be a promising line of future research.

References

[1] S. Agarwal, R. Agrawal, P.M. Deshpande, A. Gupta, J.F. Naughton, R. Ramakrishnan, and S. Sarawagi. On the computation of multidimensional aggregates. In *Proc. 1996 Int. Conf. Very Large Data Bases (VLDB'96)*, Bombay, India, 1996.

[2] S. Chaudhuri and U. Dayal. An overview of data warehousing and OLAP technology. *SIGMOD Record*, 26: 65–74, Toronto, Canada, 1994.

[3] S. Chawathe, V. Krishnamurthy, S. Ramachandran, and S. Sarma. Managing RFID data. In *Proc. Intl. Conf. on Very Large Databases (VLDB'04)*.

[4] K. Finkenzeller. *RFID Handbook: Fundamentals and Applications in Contactless Smart Cards and Identification*. John Wiley & Sons, 2003.

[5] C. Floerkemeier and M. Lampe. Issues with RFID usage in ubiquitous computing applications. In *Pervasive Computing (PERVASIVE) Lecture Notes in Computer Science*, 3001: 188–193, 2004.

[6] E. Fredkin. Trie memory. In *Communications of the ACM*, Vol. 3, Issue 9, 490–499, September 1960.

[7] "EPCglobal Inc, EPC radio-frequency identity protocols class-1 generation-2 UHF RFID procol for communications at 860 MHz to 960 MHz." http://www.epcglo balinc.org/standards_technology/EPCglobal2UHFRFIDProtocolV109122005. pdf (2003).

[8] H. Gonzalez, J. Han, and X. Li. Flowcube: Constructuing RFID flowcubes for multidimensional analysis of commodity flows. In *Proc. 2006 Int. Conf. Very Large Data Bases (VLDB'06)*, Seoul, Korea, September 2006a.

[9] H. Gonzalez, J. Han, and X. Li. Mining compressed commodity workflows from massive RFID data sets. In *Proc. 2006 Conf. on Information and Knowledge Management (CIKM'06)*, Arlington, VA, November 2006b.

[10] H. Gonzalez, J. Han, X. Li, and D. Klabjan. Warehousing and analysis of massive RFID data sets. In *Proc. 2006 Int. Conf. Data Engineering (ICDE'06)*, Atlanta, GA, April 2006c.

[11] H. Gonzalez, J. Han, and X. Shen. Cost-conscious cleaning of massive RFID datasets. In *Proc. 2007 Int. Conf. Data Engineering (ICDE'07)*, Istanbul, Turkey, April 2007.

[12] J. Gray, S. Chaudhuri, A. Bosworth, A. Layman, D. Reichart, M. Venkatrao, F. Pellow, and H. Pirahesh. Data cube: A relational aggregation operator generalizing group-by, cross-tab and sub-totals. *Data Mining and Knowledge Discovery*, 1: 29–54, 1997.

[13] V. Harinarayan, A. Rajaraman, and J.D. Ullman. Implementing data cubes efficiently. In *Proc. 1996 ACM-SIGMOD Int. Conf. Management of Data (SIGMOD'96)*, Montreal, Canada, 1996.

[14] "MIT Auto-ID Center, 13.56 MHz ism band class 1 radio frequency identification tag interface specification. Technical report. http://www.epcglobalinc.org/ standards_technology/Secure/v1.0/VHF-class0.pdf (2003).

[15] S.R. Jeffrey, M. Garofalakis, and M.J. Franklin. Adaptive cleaning for RFID data streams. In *Proc. 2006 Int. Conf. Very Large Data Bases (VLDB'06)*, Seoul, Korea, September 2006a.

[16] S.R. Jeffrey, G. Alonso, M. Franklin, W. Hong, and J. Widom. A pipelined framework for online cleaning of sensor data streams. In *Proc. 2006 Int. Conf. Data Engineering (ICDE'06)*, Atlanta, GA, April 2006b.

[17] S. Sarma, D.L. Brock, and K. Ashton. The networked physical world. White paper, MIT Auto-ID Center, http://archive.epcglobalinc.org/publishedresearch/MIT-AUTOID-WH-001.pdf (2000).

[18] S.E. Sarma, S.A. Weis, and D.W. Engels. RFID systems, security & privacy implications. White paper, MIT Auto-ID Center, http://archive.epcglobalinc.org/ publishedresearch/MIT-AUTOID-WH-014.pdf (2002).

[19] A. Shukla, P. Deshpande, and J.F. Naughton. Materialized view selection for multidimensional datasets. In *Proc. 1998 Int. Conf. Very Large Data Bases (VLDB'98)*, New York, 1998.

Chapter 4

RFID Data Management: Issues, Solutions, and Directions

Quan Z. Sheng, Kerry L. Taylor, Zakaria Maamar, and Paul Brebner

Contents

Radio Frequency IDentification (RFID) technology has existed for the past 50 years. However, only in recent years, has RFID technology moved from obscurity into mainstream applications for automatic identification. The driving force behind this major shift is the increasing affordability of RFID tags, the convergence of industrial standards, the added value offered to applications, and the efforts that EPCglobal and industry giants like Wal-Mart have put into the use of RFID technology. While RFID technology provides promising benefits, such as inventory visibility and business process automation, several significant challenges, such as data processing and management, integration architecture design, security, and privacy need to be seriously addressed before these benefits can be fully realized. In this chapter, we survey the main techniques for RFID data processing and integration, a central concern in applying RFID in real-world applications, and propose a set of criteria for assessing the existing RFID techniques and products. Open research issues for RFID data processing and integration are also discussed.

4.1 Introduction

Uniquely identifying individual objects is a powerful capability, which is essential to many aspects of modern life, such as manufacturing, distribution logistics, access control, and fighting terrorism. Today's identification systems consist of identifying markings (e.g., barcodes, photos, signatures) and ways to read them (e.g., scanners, photo detectors, humans). RFID is a wireless communication technology that is useful in identifying objects accurately [19,40,68]. RFID uses radio-frequency waves to transfer data between tagged objects and readers without line of sight, providing an excellent support for automatic identification.

Although RFID has been around for more than a half century [37], it is only in recent years that this technology has been gaining significant momentum due to the convergence of lower cost and increased capabilities of RFID tags. Currently, RFID is emerging as an important technology for revolutionizing a wide range of applications, including supply chain management, retail, aircraft maintenance, anti-counterfeiting, baggage handling, and healthcare [7,40,55,69]. It also heralds the emergence of inexpensive and highly effective pervasive computers that will have dramatic impacts on individuals, organizations, and societies [59]. Many organizations are planning or have already exploited RFID in their main operations to take advantage of the potential of more automation, efficient business processes, and inventory visibility. For example, a recent news shows that Wal-Mart has reduced out-of-stocks by 30 percent on average after launching its RFID program [33]. Many predictions agree that RFID will be worth billions of dollars in new investments [3]. According to IDTechEx, a leading market research and advisory firm, the RFID market will rocket from $2.71 billion in 2006 to $26.23 billion in 2016 [13].

While RFID provides promising benefits in many applications, some significant challenges need to be overcome before these benefits can be realized. An important challenge in RFID applications is *data management*. This has been a central concern because RFID data possess significant characteristics—such as streaming and large volume, noisy, spatial and temporal, implicit semantics—that require special attention [9,65]. The deployment of large-scale RFID applications also offers unique challenges in terms of scalability, heterogeneity, manageability, and legacy systems. Last but not least, given the ability of inexpensively tagging and, thus, monitoring a large number of items (e.g., people), RFID raises some serious security and privacy concerns that are critical issues in adapting RFID technology [24,43].

Many researchers are currently engaged in developing solutions for these challenges. In this chapter, we present a survey on the main issues and solutions that hinder and boost the exploitation of RFID technology, respectively. The aim of this work is to provide a better understanding of the current research issues and activities in this field. However, it is worth mentioning that we do not cover, every possible aspect of RFID technology. For example, hardware research (tags design, reader interference study) is important for RFID, but is out of the scope of this chapter because the focus is on the techniques for using RFID in advanced business applications, such as real-time product tracking and tracing. Although some surveys about RFID have been conducted recently (e.g., [2,12,21,34,48,51,69]), they are mostly fragmented and lack a holistic view of the problem. In particular, the work covered in those surveys did not specifically focus on RFID data processing and integration. In this chapter, we propose a framework to compare techniques for developing RFID applications. The framework identifies three layers of RFID systems: RFID hardware, RFID data processing, and RFID data integration. It also proposes a set of dimensions, which are used as a benchmark, to study these RFID solutions.

To the best of our knowledge, this is the first effort that studies the state-of-the-art techniques on RFID data management and application development. Thus, the chapter is valuable to readers looking for a general overview of the topic. It is also useful to professional RFID users seeking for guidance in the development of RFID applications.

The remainder of the chapter is organized as follows. In section 4.2, we first define the different layers in generic RFID systems and identify a set of dimensions for comparing RFID solutions across these layers. In section 4.3 and section 4.4, we provide a detailed investigation of current proposals for RFID data processing and integration. These approaches and platforms are then evaluated against a set of dimensions. We also highlight some likely future directions for research and development. Finally, in section 4.5, is the conclusion.

4.2 Overview of RFID Frameworks

RFID is emerging as a promising technology for a wide range of applications, such as access control, supply chain management, and healthcare. However, most existing RFID applications are still in their early stages and have been designed in a fairly ad-hoc manner. There is no standard for RFID application development as of yet and there is a lack of definitions of requirements that RFID systems must satisfy. In this section, we first present a typical architecture of an RFID system and identify the different layers that make up such a system. We then identify a set of dimensions, which will be used as a benchmark, to study RFID data processing and integration issues in the development of RFID applications.

4.2.1 Architecture of an RFID System

RFID-enabled applications refer to the use of RFID technology and computerized systems (e.g., networking services, databases, Web servers) for achieving advanced business goals like real-time products tracking, automatic warehouse management, just to cite a few [40]. It is of interest to note that deploying an RFID application by directly hooking RFID readers to backend systems is considered as disastrous as watering a lawn by hooking a garden hose directly to a fire hydrant [45]. An appropriate architecture is the key to the success of RFID applications. The building blocks in RFID applications are provided through an RFID framework (Figure 4.1). Typically, the purpose of an RFID framework is to enable data to be: (1) transmitted by a portable device (i.e., an RFID tag), which is read by an RFID reader; (2) processed; and (3) adapted according to the needs of a particular application.

Figure 4.1 depicts the main components of a generic RFID system. Interactions in such a system occur in three layers: devices, data processing, and data integration. The RFID devices layer consists of RFID tags and readers as well as RFID protocols (e.g., ISO 14443, electronic product code (EPC) Class 0 and 1) for reading and writing RFID data. The RFID data processing layer consists of a number of software components for communicating with RFID readers, filtering and cleaning RFID data, and adapting RFID data for high-level applications, including semantic filtering and automatic data transformation and aggregation. RFID data can be formatted using, for example, the Physical Markup Language (PML) [20] and sent to different targets as messages, streams, or other formats via Web services: JMS, HTTP response, or TCP/IP data packets.

The RFID data integration layer is concerned with the applications that exploit RFID data, such as supply chain management. RFID data could be both local RFID data and the one from its business partners, which is typically the case when applications involve multiple business entities. This requires that the applications understand the semantics of RFID data that is possibly disparately represented. The objective of interactions at this layer is to achieve a seamless weaving of RFID data and business processes.

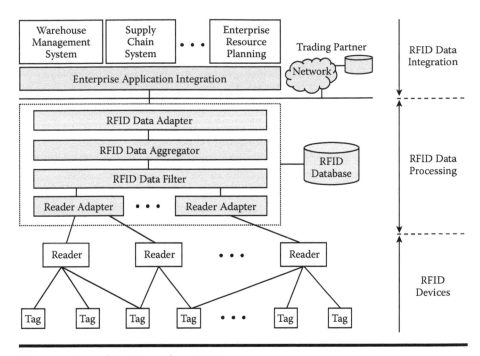

Figure 4.1 Architecture of an RFID system.

This chapter focuses on the part of the system that lies above the RFID devices layer. (Readers are referred to [10,15,19,21,55,67] for more details on physical aspects related to RFID devices and relevant research problems.)

4.2.2 Dimensions for RFID Data Processing

RFID data possesses significant characteristics [9,65], that have to be fully considered by RFID data processing approaches. Therefore, the characteristics of RFID data are taken as the dimensions for evaluating RFID data processing approaches (i.e., whether or not an approach handles a particular characteristic of RFID data).

Despite the diversity of RFID applications, RFID data shares some common fundamental characteristics:

■ *Inaccurate*: While the accuracy of current RFID readers is improving, there are still erroneous readings in RFID systems, such as duplicate reads, missed reads, and ghost reads [14], due to interference, temporary or permanent malfunction of some components. Therefore, RFID data tends to be noisy and unreliable.

■ *Streaming and large volume*: RFID data is quickly and automatically generated. The data volume generated can be enormous. For example, Venture Development Corporation [64] predicts that, when tags are used at the item level, Wal-Mart will generate around 7 terabytes of data every day.

■ *Dynamic and temporal*: The RFID data is dynamically generated and associated with timestamps when the readings are made. All RFID-related transactions are, in fact, associated with time. Temporal information is important for tracking and monitoring RFID objects. For example, with timestamps, it is possible to get the information on how long it takes for an aircraft part to move from the warehouse to the maintenance venue.

■ *Location-based*: Tagged objects are typically mobile by going through different locations during their life cycles. For example, in supply chain systems, a product travels among different physical sites, such as manufacturer, distribution center, retailer, and finally customer.

■ *Implicit inferences*: RFID data always carries implicit information, such as changes in states and business processes, and further derived information like aggregations (e.g., containment relationship among RFID objects). In order to be applied, RFID data always requires the context of other information, such as business applications, environmental situations, etc. For example, if a case of cameras and a pallet are captured together in a packing station, a desirable inference could be that the case has been packed in the pallet.

■ *Limited active life span*: RFID data normally has a limited active life period, during which the data is actively updated, tracked, and monitored. For example, in a supply chain system, this period starts when the products are delivered from the manufacturer and ends when the products are sold to customers. Life-cycle management of RFID data is of paramount importance to the performance of RFID-enabled applications due to a large volume of RFID data.

4.2.3 Dimensions for RFID Data Integration

Integration of RFID data into business processes presents many challenges [6]. The development of RFID applications has to consider the usage scenarios, business requirements, and even the parties involved. There are specific tradeoffs with regard to the requirements of RFID applications. Thus, it is important to determine the relevant requirements and understand the related tradeoffs when evaluating RFID systems. The following dimensions as considered for evaluating RFID integration infrastructures:

■ *Scalability*: Scalability refers to the ability of a system to grow in one or more dimensions, such as the volume of RFID data and the number of transactions, without affecting the system's performance. With the adoption of RFID technology, companies are required to handle RFID data from thousands of readers distributed across various sites. More importantly, changes in the business climate are forcing companies to ally with other business partners in order to be competitive in the global market. Thus, scalability is an important criterion to consider when evaluating integration solutions in RFID systems.

■ *Heterogeneity*: The deployment of an RFID solution may be distributed across sites, companies, or even countries, where different RFID hardwares, data structures, and standards may be used. This naturally requires that RFID systems support the distribution of message preprocessing functionality (e.g., filtering and aggregation), as well as business logic across multiple nodes to better map to existing company and cross-company structures [6].

■ *Manageability*: This dimension refers to the degree of the visibility and manageability of RFID systems. For the successful deployment of an RFID solution in large-scale, distributed applications, good administration and testing support are prerequisites. RFID systems must facilitate the supervision, testing, and control of their individual components as well as end-to-end processing of RFID data.

■ *Security*: Vast amounts of potentially sensitive data involved makes security a critical issue in RFID systems [23,51,69]. For instance, an organization that uses RFID in its supply chain does not want competitors to track its shipments and inventory. Furthermore, standard security mechanisms, for example, SSL (secure socket layer), are resource consumption intensive and are not suitable for RFID systems where other CPU-intensive work (e.g., filtering) is involved. In Rieback et al. (2006a) [50], a self-replicating RFID virus was discussed. It uses RFID tags as a vector to compromise backend RFID systems via an Structured Query Language (SQL) injection attack. In J. Westhues [71], cloning an RFID tag is demonstrated. Before RFID technology reaches its real potential, sophisticated security measures must be in place to boost RFID participants confidence that their applications are safely protected.

■ *Flexibility*: This dimension refers to the degree of adaptability of a system. RFID systems need to be adaptable to different business scenarios and their changes. Changes may be introduced to adapt RFID applications to actual business climate (e.g., economic, policy, or organizational changes), or to take advantage of new business opportunities. In addition, RFID systems must provide flexible means to rapidly respond to abnormal situations (e.g., missing of expected goods).

■ *Openness*: Openness refers to the interoperability of RFID systems. For instance, a well-designed reader adapter at the edge makes RFID integration reader-agnostic. In addition to being hardware-agnostic, RFID systems should be based on existing communication protocols (e.g., TCP/IP, HTTP), as well as syntax and semantics standards (e.g., XML, PML [20], EPC [16]). An open RFID architecture will allow the use of RFID devices from a wide array of hardware providers and, more importantly, will support the deployment of RFID solutions across institutional or country boundaries.

4.3 State-of-the-Art Technologies for RFID Data Management

Due to the distinguished features of RFID data (Section 4.2.2), it is useless for applications to directly use raw RFID data. Data must be appropriately processed before it can be used in an application. Fortunately, the database research community is developing a strong interest in RFID data management [6,9,25,26,28,30,31,49,65,66]. In this section, we overview a set of representative research work on RFID data management.

4.3.1 Siemens RFID Middleware

Siemens RFID Middleware [65,66] is an integrated RFID data management system that: (1) enables semantic RFID data filtering and automatic data transformation, (2) provides powerful query support of RFID object monitoring and tracking, and (3) is easy to integrate into other applications. An expressive temporal-based data model called DRER (Dynamic Relationship ER Model) is proposed for RFID data. DRER abstracts a set of entities, namely Object, Reader, Location, and Transaction (Figure 4.2). The interactions between entities are modeled as relationships. Two types of relationships are identified in DRER: (1) state-based relationship that generates state history, and (2) event-based relationship that generates events. For the state-based relationship, two attributes (`tstart` and `tend`) are associated with a relationship to represent the lifespan of the relationship (e.g., the period that a product stays in the warehouse). For the event-based relationship, an attribute `timestamp` is associated with a relationship to indicate the time of the event

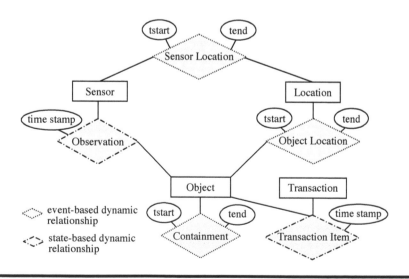

Figure 4.2 Dynamic Relationship ER (DRER) Model.

occurrence. All DRER entities and relationships can be implemented as tables in a Relational Database Management System (RDBMS). Because of the maintenance of temporal and location history of RFID objects, it becomes easy to track and monitor RFID objects (e.g., missing object searching).

To provide support of automatic RFID data transformation between physical and virtual worlds, a declarative rule-based, event-oriented approach is also proposed where RFID application logic is devised into complex events. Siemens RFID Middleware developed a graph-based RFID event detection engine called RCEDA (RFID complex event detection algorithm) for effective processing of highly temporally constrained RFID events.

4.3.2 ESP

ESP (extensible sensor stream processing) [30–32] is a framework for cleaning the data streams produced by physical sensor devices. A programmable pipeline of five cleaning stages is introduced in ESP, namely Point, Smooth, Merge, Arbitrate, and Virtualize (Figure 4.3). These stages operate on different aspects of the data, from finest (single reading) to coarsest (readings from multiple sensors and other data sources). The Point stage operates over a single value in a sensor stream to filter individual values (e.g., errant RFID tags or outliers) or convert fields within an individual tuple. The Smooth stage uses the temporal granule defined by the application to correct missed readings and detect outliers over a sliding window of a single sensor stream, whereas the Merge stage uses the application's spatial granule for corrections and detections over sensor streams within a proximity group. Finally, the

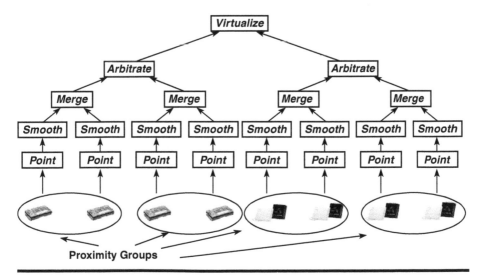

Figure 4.3 ESP processing stages.

Arbitrate stage deals with conflicts (e.g., duplicate readings) between data streams from different spatial granules, and the Virtualize stage combines readings from different types of devices and different spatial granules.

The approach, however, is mainly based on the assumption of the temporal and spatial correlations among sensed data. Therefore, it is only suitable for applications that are not interested in individual readings, but rather in temporal or spatial granules, such as the redwood tree environmental monitoring application [63].

4.3.3 Oracle's EPC Bitmap

The work from Oracle [28] proposes an EPC bitmap datatype for efficiently handling high volume data generated by RFID-based, item-level tracking applications. An EPC bitmap datatype is defined to represent a collection of EPCs, which share common EPC segments (e.g., header, manager number, and object class). The key observation underpinning the approach is that RFID tagged items, in most applications, can be tracked in groups based on common properties, such as location, expiration date, or manufacturer. Collections of EPCs can be represented as an EPC bitmap, which can be accessed and manipulated using a set of EPC bitmap operations (e.g., *epc2bmp* and *bmp2epc* for conversion). The key benefits of introducing the bitmap datatype include: (1) it compactly represents a collection of RFID identifiers without losing information, and (2) the typical operations on a pair of collections can simply be performed using bitmap operations on the corresponding bitmaps. As a consequence, significant storage savings can be achieved, while the query performance remains the same or better than traditional approach (i.e., collections of EPCs are stored natively in a Database Management System (DBMS)).

4.3.4 RFID Warehouse

The work from the University of Illinois at Urbana-Champaign (UIUC) [25,26] proposes a novel model for warehousing RFID data that allows efficient and flexible high-level, OLAP (online analytical processing)- style analysis in multidimensional space. The model consists of a hierarchy of highly compact summaries, represented as RFID-Cuboids, of the RFID data aggregated at different abstraction levels where data analysis takes place. Each RFID-Cuboid records object movements and consists of three tables: (1) *information* table that stores product information for each RFID object, (2) *stay* table that stores information on objects that stay together at a location, and (3) *map* table that stores path information necessary to link multiple stay records. The basic principles of data structure design take advantage of the fact that individual objects tend to move and stay together so that multiple movements of RFID objects can be collapsed into a single record without loss of information. The proposed model results in orders of magnitude of smaller database size as well as efficient support of a wide range of RFID queries.

4.3.5 Evaluations and Open Issues

The aforementioned research projects are compared using the RFID data processing dimensions presented in section 4.2.2. Table 4.1 summarizes the results. For
example, Siemens RFID Middleware partially addresses the inaccuracy issues of
RFID data by supporting the filtering of duplicated readings. The work from Oracle addresses explosion of RFID data using EPC bitmap datatype.

From the table, we can see that most of the projects support the modeling of
temporal and spatial characteristics of RFID data. It is also obvious that handling
large volumes of RFID data is a popular trend. However, it is worth mentioning
that inaccuracy, implicit references, and active lifespan are neglected by most projects. Only Siemens RFID Middleware provides very limited support.

Although a number of efforts and results have been made and obtained in this
area, RFID data processing technology is still far from mature and requires significant efforts to address some open research challenges:

- *Accurate and reliable RFID data provisioning*: The unreliability of RFID
  data has been widely considered as a principle challenge in RFID applications [9,31]. The accuracy of RFID data is extremely important for RFID-
  enabled applications in the sense that RFID data is mainly used to automate
  business processes and applications. The imperfection of RFID information
  may misguide users using RFID applications. For example, a ghost read at
  a checkout point might trigger a charge to the customer although she is not
  purchasing the corresponding goods. We believe that extensive research is
  needed for highly reliable provisioning of RFID data.
- *General approach for efficient RFID data management*: RFID systems may
  generate large volume of data [9,64]. Accumulation of RFID data can easily
  lead to poor performance (e.g., slower queries and updates). Therefore, efficient management of large amount of RFID data is a critical issue that must
  be dealt with in RFID applications. Current solutions on efficient RFID
  data management focus on data mining-based approaches [26,28] that
  assume RFID data shares some common properties (e.g., move together in
  bulky mode, with the same expired date) and can be grouped based on such
  properties, thereby fitting well to only a limited number of RFID applications (e.g., supply chain management). A more general approach, therefore,
  is needed for handling massive RFID data. Given the limited active lifespan of RFID data, we believe that novel mechanisms (e.g., partitioning of
  RFID data) will become increasingly important.
- *Flexible and intelligent RFID data transformation and aggregation*: Raw
  RFID data presents little value without transforming into more meaningful form that is suitable for application-level interactions. Moreover, RFID
  data carries implicit meanings and associated relationships (e.g., containment) with other RFID data, where appropriate inferences have to be

Table 4.1 Comparison: Research Projects Versus RFID Data Handling Requirements

	Inaccurate	Large Volume	Temporal	Spatial	Implicit References	Active Lifespan
Siemens RFID Middleware	Partial addressed: filtering duplicated readings	RFID data partitioning	DRER (Dynamic Relationship ER) model	DRER model	Rule-based data transformation	Considered in RFID data partitioning
Berkeley's ESP	ESP (extensible sensor stream processing) scheme for data cleaning	Not addressed	Temporal granules	Spatial granules	Not addressed	Not addressed
Oracle's EPC bitmap	Not addressed	EPC bitmap datatype for a collection of EPCs	Not explicitly addressed	Not explicitly addressed	Not addressed	Not addressed
UIUC's RFID Warehouse	Not addressed	RFID-Cuboid data structure	Stay table	Stay table	Not addressed	Not explicitly addressed

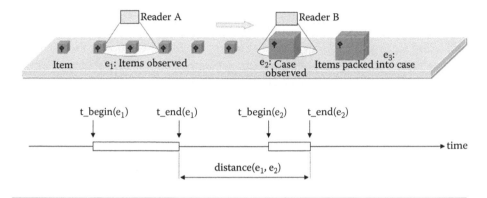

Figure 4.4 Complex events for RFID data aggregation.

made [9]. For example, in Figure 4.4, two observations of the items and a case happening in a certain time interval may imply that the items are packed into the case. This normally requires the formation of relationships among RFID objects and the devising of business logics of RFID applications, which has been identified as a difficult issue for RFID applications [44]. The extension of event modeling techniques is a natural step in this direction. Indeed, a few research efforts for event-driven RFID data management are emerging [22,66].

4.4 RFID Integration Platforms

Major software vendors like Sun Microsystems, SAP, BEA Systems, and IBM, and research organizations, such as WINMEC at UCLA, Auto-ID labs, and Siemens Research are currently working on implementing RFID integration platforms. Due to the large number of products, both existing and emerging, we will not cover all of them in this chapter. Instead, we focus on the major players in this arena and overview the main features of a set of representative platforms. It should be noted that for commercial products, our judgment is based on user manuals and white papers because there are few or no technical publications detailing these products.

4.4.1 UCLA WinRFID

WinRFID [47] is a multilayered middleware system developed using Microsoft® .NET [39] framework, that enables rapid RFID application development. It consists of five main layers: physical, protocol, data processing, XML framework, and data presentation. The physical layer deals with the hardware, including readers, tags, and other sensors. The protocol layer abstracts the reader-tag protocols. The data processing layer processes the data streams generated by the RFID reader network. Finally, XML framework and data presentation layers deal with data representation and presentation, respectively. Middleware modules are deployed as independent components that run in dispersed machines. Currently, a Web service-based distributed architecture is under development.

To support the extendibility of its platform, WinRFID exploits the runtime plug-in feature of .NET. New readers, new protocols, and new data transformation rules can be added to corresponding modules with minimum disruption of the existing infrastructure. Data filtering, aggregation, and adaptation are handled by a rule-based engine that is tightly coupled in the architecture of WinRFID.

4.4.2 Siemens RFID Middleware

The Siemens RFID Middleware [65,66] is a RFID data management system, which provides an integrated solution for RFID applications. It offers the following functionalities: automatic data acquisition, filtering, and transformation; expressive data modeling and effective query support; and adaptation to different applications with minimum configurations.

The Siemens RFID Middleware consists of three main components: RFID readers, Event managers, and RFID data server. Event managers dynamically receive data from readers, preliminarily filter the data, and forward the data (formatted in PML) to the RFID data server. The RFID data server includes: (1) RFID data manager, for RFID data processing (i.e., semantic data filtering and data aggregation), RFID object tracking and monitoring, and decision-making support; (2) RFID data store for RFID data maintenance (i.e., storing, partitioning, and retrieval of RFID data); and (3) application integration interface for integrating system with other applications.

4.4.3 SAP Auto-ID Infrastructure

SAP Auto-ID Infrastructure (SAP AII) [6,53] is a middleware that adds a layer of intelligence between automated communication and sensing devices (e.g., RFID readers, Bluetooth devices, and barcode devices) and enterprise applications. SAP AII is built on SAP's Web Application Server, which is part of the company's NetWeaver [54] integration and application platform. It converts RFID and sensor data into business process information by associating the data with specified mapping rules (e.g., forwarding observation data directly to backend systems, executing predefined business logic) and metadata.

The architecture of SAP AII consists of four system layers: device layer, device operation layer, business process bridging layer, and enterprise application layer. The device layer supports different types of sensor devices via a hardware-independent, low-level interface. The device operation layer contains one or more device controllers, coordinating multiple devices. The business process bridging layer associates observation messages with existing business processes, realized by a component called Auto-ID node. There can be multiple Auto-ID nodes in SAP AII. Finally, the enterprise application layer supports business processes of enterprise applications, such as supply chain management, customer relationship management, running on either SAP or non-SAP backend systems.

To allow the testing of an Auto-ID deployment without the installation of physical devices, SAP AII also provides tools for simulating RFID readers and messages. To support the flexibility of the infrastructure, SAP AII exploits a customizable rule-based approach in specifying and executing business logic at the Auto-ID nodes.

4.4.4 Sun Java System RFID Software

Java System RFID Software [61] is one of the first RFID integration platforms focusing on large-scale deployments. It is designed on a service-oriented architecture, providing network services to applications through a number of standard protocols and interfaces. The bottom of the RFID software architecture are tag readers or sensors that are responsible for reading tagged items. The data is sent to Java System RFID Software for processing. The topmost layer of the architecture is enterprise information systems to be integrated, such as enterprise resource planning.

Java System RFID Software consists of two major components: the Java System RFID event manager and the Java System RFID information server. The Java System RFID event manager is designed to process (filter and aggregate) RFID data. It implements a federated service architecture, which essentially provides distributed, self-organizing, and network-centric capabilities. This architecture makes the RFID event manager not only scalable, but also adaptable to unforeseen changes on the network. The Java System RFID information server provides access to the business events generated by the RFID event manager as well as serves as an integration layer that offers options for integrating with enterprise applications.

The Java System RFID Software architecture is built with Java Enterprise System [60], which delivers a set of shared technology components and services for enhanced integration and simplified maintenance. Network identity and security services from the Java Enterprise System, for instance, help secure the Java System RFID Software.

4.4.5 EPCglobal Network

The EPCglobal Network is an architecture that is often referred to in literature as the "Internet of things" [56]. It has been designed by the Auto-ID Center and developed further by the EPCglobal. The basic idea behind the framework is to realize a so-called data-on-network system, where RFID tags contain an unambiguous ID, namely Electronic Product Code (EPC), and other data pertaining to the objects is stored and accessed over the Internet.

The major components of the EPCglobal Network include the Object Naming Service (ONS) and the EPC Information Service (EPCIS). The ONS provides reference to the saved object information through EPC, thus helping the discovery of RFID objects over the network. EPCIS offers an interface for accessing the stored RFID data. To exchange RFID data between EPCglobal network and external applications, an XML-based markup language, PML, is proposed. The advantages of EPCglobal network include the possibility of using low-cost RFID tags and a more scalable architecture. It is worth mentioning that in order to track RFID objects in the EPCglobal network, the movement of these objects has to be continuously published to a centralized discovery service, which unfortunately, prevents the network from realizing a full-fledged distributed architecture [1].

4.4.6 *Other Platforms*

BEA WegLogic RFID. BEA WebLogic RFID products [5] consists of BEA WebLogic RFID Edge Server, BEA WebLogic RFID Enterprise Server, and BEA WebLogic RFID Compliance Express. RFID Edge Server is a lightweight software that can be deployed at various sites. It provides a comprehensive software infrastructure for filtering, integrating, disseminating RFID data, and monitoring and configuring RFID readers. RFID Enterprise Server provides the standards-based infrastructure for centrally managing RFID data collected from RFID Edge Servers. Finally, RFID Compliance Express is designed to meet current compliance challenges while establishing a foundation for future expansion.

IBM WebSphere RFID. The IBM RFID solution [29] is made up of three components: devices, WebSphere RFID Premises Server, and a WebSphere integration server. RFID devices (e.g., readers, scanners, printers) have to be embedded with the WebSphere RFID Device Infrastructure, which is a software that supports functions for RFID data collection and delivery. WebSphere RFID Premises Server is a middleware that aggregates, monitors, interprets, and escalates RFID data, and provides interface for the integration (via an integration server, such as WebSphere Business Integration Server) with enterprise applications.

Sybase RFID Solutions. The core part of the Sybase RFID solutions [62] is a layered RFID architecture, which consists of six major components: the physical device layer, the RFID network layer, the process layer, the persistence layer, the integration layer, and the presentation layer. The RFID network layer is implemented through Sybase RFID Anywhere, a middleware platform that supports the reading, writing, filtering, grouping, and routing of RFID data generated by readers and related peripherals. The architecture also provides tools for data analysis and mobilization (e.g., delivering RFID data to mobile devices).

4.4.7 Evaluations and Open Issues

In Table 4.2, major RFID integration platforms are compared using the RFID data integration dimensions (section 4.2.3). For example, scalability of WinRFID is supported by deploying RFID middleware components as self-contained, distributed modules. Heterogeneity is partially addressed through the abstraction of published protocols and the usage of XML framework for data representation. Openness is provided by exploiting a standard-based framework, such as EPC, Web services, and XML. Flexibility is supported by a rule-based engine and a plug-ins framework based on .Net platform. Manageability is provided by WinRFID management console. Finally, security is partially addressed through authentication and access restriction support of RFID data.

From the table, one can see that most RFID integration platforms pay particular attentions to the issues like scalability, heterogeneity, and openness in their design. However, they either totally neglect or provide very limited support to the issues like flexibility, security, and privacy. Although the current solutions provide the foundation for building RFID applications, several research issues still need to be addressed. In particular, we identify the following directions for future research, namely support of large scale applications, seamless integration of legacy systems and environmental sensors, security, and privacy.

■ *Support of large scale applications*: In spite of many small RFID pilot applications (e.g., warehouse monitoring), there present significant challenges for developing and deploying large-scale and distributed applications (e.g., supply chain management across companies). First, the nature of RFID applications as well as scalability requirements call for a completely distributed system architecture. Although some of the current existing platforms support the distribution of some functionalities (e.g., RFID edge processing), a full-fledged distributed architecture is needed where functionalities can be distributed and replicated, and data can be shared and synchronized across multiple nodes. Second, the deployment of a large distributed system requires the ability to continuously monitor the state of the system and adaptively adjust its behavior. Given the large number and highly distributed nature of RFID devices, the administration tools for visualizing, configuring, testing, and monitoring RFID devices and system components will become increasingly important.

Table 4.2 Comparison: Integration Platforms Versus RFID Data Integration Requirements

	Scalability	Heterogeneity	Manageability	Security	Flexibility	Openness
WinRFID	Self-contained, distributed middleware modules	Abstracted protocol module for published protocols and XML framework for data representation	WinRFID management console	Authentication and access restriction support of RFID data	Rule-based engine, WinRFID plug-ins	Standard-based framework (e.g., Web services, XML)
Siemens RFID Middleware	Not addressed	Reader adapters for RFID readers and PML for formatting RFID data	Not addressed	Not addressed	Rule-based framework	Standard-based framework, using ONS and XML
SAP AII	Distributed device controllers and Auto-ID nodes	Hardware-independent device interface, distributed Auto-ID nodes	Auto-ID administrator, tools for simulating and testing RFID messages and readers	Not addressed	Customizable rule-based engine	Compliant with the standards proposed by the EPCglobal consortium
Java System RFID Software	Distributed architecture of the RFID event manager	Extensible device adapters	A browser-based interface for centralized monitoring and management of devices and services	Security services in Java Enterprise System	Not addressed	Standard-based, service-oriented architecture (SOA)

EPCglobal Network	Distributed EPCIS, centralized discovery service	EPC-compliant devices	Not explicitly supported	Under development	Not addressed	Standard-based framework, such as Web service and XML-based PML
WebLogic RFID Product Family	Lightweight, distributed architecture for RFID Edge Server	Out-of-box support of major RFID readers; simple object access protocol (SOAP) interfaces	Administration Console; Monitoring and management agent; reader simulator	Not addressed	RFID Compliance Express	Standards compliance (e.g., EPCglobal and ISO standards), SOA
WebSphere RFID Server	WebSphere RFID Device Infrastructure (only devices embedded with the infrastructure)	Not addressed	Offer system management capabilities, such as management and monitoring of hardwares and applications in remote locations	Not addressed	Not addressed	Standard-based framework
Sybase RFID Architecture	Distributed RFID edge processing	Strategic relationships with RFID hardware vendors, out-of-box integration capabilities of existing and emerging infrastructures	Interface for device management	Role-based authorization	Not addressed	Support of open standards (e.g., ISO and EPC standards), SOA

■ *Seamless integration of legacy systems and environmental sensors*: Support of integration of existing systems is of paramount importance for RFID integration platforms. These legacy software systems are normally matured, heavily used, and constitute massive corporate assets [57]. It is costly, or impossible, to scrap the legacy systems and replace them with new ones. Many businesses seek ways to maximize their existing investment and preserve valuable business knowledge while adapting to rapidly evolving technologies. Currently, vendors and researchers are developing RFID middlewares (e.g., SAP AII [53], Sybase RFID [62], and WinRFID [47]) that will link new RFID systems into existing infrastructures. We expect more research effort in this direction. Moreover, many researchers believe that integration of sensors and RFID tags will make it ideal for obtaining information about the physical world [40,7]. An RFID tag could return environmental data along with an object's identity. For example, a sensor-enabled RFID tag attached to an airplane part could record the stress and shock experienced during a flight, which could be used by a maintenance officer to make the preventive maintenance schedule for this particular part. Unfortunately, RFID readers are discrete event sources, whereas environmental sensors provide a stream of periodic events. This mismatch has to be represented and resolved in order to seamlessly integrate RFID and environmental sensors [6,58].

■ *Security*: RFID systems are increasingly being used in high security applications, such as access control [27], electronic passports [72,4], and systems for making payments [17,8,18] or issuing tickets [38]. Security is a critical issue for RFID technologies to be adopted and must be enforced to give businesses the confidence that their data is safely handled. A few solutions have proposed authenticating and encrypting mechanisms between RFID tags and readers [34,51]. However, data encryption and authentication increase the resource consumption and the latency of read cycles [23,46]. Intensive work is needed for developing comprehensive solutions that not only protect RFID information, but maintain desirable system performance.

■ *Privacy*: Privacy is another fundamental concern with RFID systems. It is generally (mis)perceived as an issue whose natural solution consists of good security mechanisms. Although security and privacy are two tightly interrelated issues, securing RFID applications does not necessarily ensure privacy. There are two notable privacy issues: (1) leaking information pertaining to personal property, and (2) tracking the consumer's spending history and patterns and physical whereabouts [43,24]. Examples of RFID privacy threads include terrorists scanning digital passports to target specific nationalities, and police abusing a convenient new means of cradle-to-grave surveillance. To address privacy threads of RFID users, not only technical solutions (e.g., Blocker Tag [35], Hash-Lock Scheme [70], Internal re-encryption scheme [36]), but also social and legal countermeasures (e.g., RFID policies [11,52], RFID bills [42,41]) need to be implemented to ensure consumers that their data will not be misappropriated [43,23].

4.5 Conclusion

The ability of RFID technology to precisely identify objects at low cost and without line of sight creates many new and exciting application areas. In the future, this wide range of applications will make RFID an integral part of our daily lives. However, despite the obvious potential, there are a number of fundamental research and development issues that persist. Development of RFID applications needs not only to deal with the unique characteristics of RFID data, but to consider specific data integration requirements of system architectures. As we have indicated, only when robust, scalable, secure solutions have been found will the promise of RFID applications be fully realized.

In recent years, the field of RFID is becoming a vibrant and rapidly expanding area of research and development. Many researchers and vendors are currently engaged in proposing data management solutions needed for the development of RFID applications. Along with the current research efforts, we encourage more insight into the problems of this nascent technology, and more development in solutions to the open research issues as described in this chapter.

Reference

[1] R. Agrawal, A. Cheung, K. Kailing, and S. Schonauer. Towards Traceability Across Sovereign, Distributed RFID Databases. In *Proc. of the 10th International Database Engineering and Applications Symposium (IDEAS'06)*, Delhi, India, December 2006.

[2] R. Angeles. RFID technologies: Supply-chain applications and implementation issues. *Information Systems Management*, 22(1):51–65, 2005.

[3] Z. Asif and M. Mandviwalla. Integrating the supply chain with RFID: A technical and business analysis. *Communications of the Association for Information Systems*, 15, March 2005.

[4] Australian Government Department of Foreign Affairs and Trade. The Australian ePassport. http://www.dfat.gov.au/dept/passports/, visited on 20/05/2006.

[5] BEA Systems. "BEA WebLogic RFID Product Family." http://www.bea.com/framework.jsp?CNT=index.htm&FP=/content/products/weblogic/rfid/, visited on 07/06/2006.

[6] C. Bornhövd, T. Lin, S. Haller, and J. Schaper. Integrating Automatic Data Acquisition with Business Processes Experiences with SAP's Auto-ID Infrastructure. In *Proc. of the 30th International Conference on Very Large Data Bases (VLDB'04)*, Toronto, Canada, September 2004.

[7] G. Borriello. RFID: Tagging the world. *Communications of the ACM*, 48(9):34–37, September 2005.

[8] J. Boyd. Here comes the wallet phone. *IEEE Spectrum*, 42(11):12–14, 2005.

[9] S.S. Chawathe, V. Krishnamurthy, S. Ramachandran, and S. Sarma. Managing RFID Data. In *Proc. of the 30th International Conference on Very Large Data Bases (VLDB'04)*, Toronto, Canada, September 2004.

[10] P.H. Cole. "Fundamentals in Radio Frequency Identification." http://autoidlabs. eleceng.adelaide.edu.au/Tutorial/SeattlePaper.doc, visited on 22/05/2006.

[11] J. Collins. "European Commission Works on RFID Policy." http://www.rfid-journal.com/article/articleview/2197/1/1/, March 2006.

[12] J. Curtin, R.J. Kauffman, and F.J. Riggins. Making the most out of RFID technology: A research agenda for the study of the adoption, usage and impact of RFID. *Information Technology and Management Journal*, 7(3-4):251–270, 2006.

[13] R. Das and P. Harrop. "RFID Forecasts, Players & Opportunities 2006–2016." http://www.idtechex.com/products/en/view.asp?productcategoryid=93, visited on 03/08/2006.

[14] D.W. Engels. On Ghost Reads in RFID Systems. Technical report, Auto-ID Center, January 2005.

[15] D.W. Engles and S.E. Sarma. The Reader Collision Problem. In *Proc. of the IEEE International Conference on Systems, Man and Cybernetics (SMC'02)*, Hammamet, Tunisia, October 2002.

[16] EPCglobal. "EPC Tag Data Standards Version 1.1." http://www.epcglobalinc.org/ standards_technology/EPCTagDataSpecification11rev124.pdf, April 2004.

[17] ExxonMobile. "Speedpass System." https://www.speedpass.com/, visited on 11/05/2006.

[18] Fastrak. "Fastrak-Keeping the Bay Area Moving." http://www.bayareafastrak.org/, visited on 22/05/2006.

[19] K. Finkenzeller. *RFID Handbook: Fundamentals and Applications in Contactless Smart Cards and Identification*. John Wiley & Sons, New York, 2003.

[20] C. Floerkemeier, D. Anarkat, T. Osinski, and M. Harrison. PML Core Specification 1.0. Technical report, Auto-ID Center, September 2003.

[21] C. Floerkemeier and M. Lampe. Issues with RFID Usage in Ubiquitous Computing Applications. In *Proc. of the 2nd International Conference on Pervasive Computing (Pervasive'04)*, Linz/Vienna, Austria, April 2004.

[22] M.J. Franklin, S.R. Jeffery, S. Krishnamurthy, F. Reiss, S. Rizvi, E. Wu, O. Cooper, A. Edakkunni, and W. Hong. Design Considerations for High Fan-in Systems: The HiFi Approach. In *Proc. of the Second Biennial Conference on Innovative Data Systems Research (CIDR'05)*, Asilomar, CA, January 2005.

[23] S. Garfinkel and B. Rosenberg. RFID: *Applications, Security, and Privacy*. Addison-Wesley Professional, Reading, MA, 2005.

[24] S.L. Garfinkel, A. Juels, and R. Pappu. RFID privacy: An overview of problems and proposed solutions. *IEEE Security & Privacy*, 3(3):34–43, May/June 2005.

[25] H. Gonzalez, J. Han, and X. Li. FlowCube: Constructing RFID FlowCubes for Multi-dimensional Analysis of Commodity Flows. In *Proc. of the 32nd International Conference on Very Large Data Bases (VLDB'06)*, Seoul, Korea, September 2006.

[26] H. Gonzalez, J. Han, X. Li, and D. Klabjan. Warehousing and Analyzing Massive RFID Data Sets. In *Proc. of the 22nd International Conference on Data Engineering (ICDE'06)*, Atlanta, GA, April 2006, visited on 11/08/2006.

[27] HID Web Site. http://www.hidcorp.com/technologies.php, visited on 11/08/2006.

[28] Y. Hu, S. Sundara, T. Chorma, and J. Srinivasan. Supporting RFID-Based Item Tracking Applications in Oracle DBMS Using a Bitmap Datatype. In *Proc. of the 31st International Conference on Very Large Data Bases (VLDB'05)*, Trondheim, Norway, September 2005.

[29] IBM. "WebSphere RFID Premises Server." http://www-306.ibm.com/software/ pervasive/ws_rfid_ premises_server/, visited on 01/07/2006.

[30] S. Jeffery, M. Garofalakis, and M. Franklin. Adaptive Cleaning for RFID Data Streams. In *Proc. of the 32nd International Conference on Very Large Data Bases (VLDB'06)*, Seoul, Korea, September 2006.

[31] S.R. Jeffery, G. Alonso, M.J. Franklin, W. Hong, and J. Widom. A Pipelined Framework for Online Cleaning of Sensor Data Streams. In *Proc. of the 22nd International Conference on Data Engineering (ICDE'06)*, Atlanta, GA, April 2006.

[32] S.R. Jeffery, G. Alonso, M.J. Franklin, W. Hong, and J. Widom. Declarative Support for Sensor Data Cleaning. In *Proc. of the 4th International Conference on Pervasive Computing (Pervasive'06)*, Dublin, Ireland, May 2006.

[33] J.R. Johnson. "Wal-Mart: RFID Reduces Out-Of-Stocks by Up to 62 Percent." http://www.dcvelocity.com/articles/rfidww/rfidww20060517/walmart_rfid.cfm, visited on 14/07/2006.

[34] A. Juels. RFID security and privacy: A research survey. *IEEE Journal on Selected Areas in Communication*, 24(2), 2006.

[35] A. Juels, R.L. Rivest, and M. Szydlo. The Blocker Tag: Selective Blocking of RFID Tags for Consumer Privacy. In *Proc. of the 10th ACM Conference on Computer and Communications Security (CCS'03)*, Washington, D.C., October 2003.

[36] S. Kinoshita, M. Ohkubo, F. Hoshino, G. Morohashi, O. Shionoiri, and A. Kanai. Privacy Enhanced Active RFID Tag. In *Proc. of 1st International Workshop on Exploiting Context Histories in Smart Environments (ECHISE'05)*, Munich, Germany, 2005.

[37] J. Landt. The history of RFID. *IEEE Potentials*, 24(4):8–11, October/November 2005.

[38] J. Libbenga. "World Cup Tickets Will Contain RFID Chips." http://www.theregister.co.uk/2005/04/04/world_cup_rfid/ visited on 20/05/2006.

[39] Microsoft. ".NET." http://www.microsoft.com/net/default.mspx, 2006.

[40] B. Nath, F. Reynolds, and R. Want. RFID technology and applications. *IEEE Pervasive Computing*, 7(1):22–24, January/March 2006.

[41] M.C. O'Connor. "California Senate Approves RFID Bill." http://www.rfidjournal. com/article/articleview/1602/1/130/, May 2005.

[42] M.C. O'Connor. "U.S. Bill Includes RFID Provision for Pets." http://www.rfidjournal.com/article/articleview/1976/1/1/, November 2005.

[43] M. Ohkubo, K. Suzuki, and S. Kinoshita. RFID privacy issues and technical challenges. *Communications of the ACM*, 48(9):66–71, September 2005.

[44] T. Palamides. "RFID 2004 Forum Report." http://www.wireless.ucla.edu/techreports2/RFID-2004-Forum.pdf, visited on 12/04/2006.

[45] M. Palmer. "Build an Effective RFID Architecture." http://www.rfidjournal.com/ article/articleview/781/1/82/, visited on 16/06/2006.

[46] T. Phillips, T. Karygiannis, and R. Kuhn. Security standards for the RFID market. *IEEE Security & Privacy*, 3(5):85–89, November/December 2005.

[47] B.S. Prabhu, X. Su, H. Ramamurthy, C.-C. Chu, and R. Gadh. WinRFID—A middleware for the enablement of radio frequency identification (RFID)-based applications. In *Mobile, Wireless and Sensor Networks: Technology, Applications and Future Directions*, 313–336. Wiley–IEEE Press, New York, March 2006.

[48] S. Pradhan, G. Lyon, I. Robertson, L. Erickson, L. Repellin, C. Brignone, M. Mesarina, B. Serra, V. Deolalikar, T. Connors, M. Jam, J. Recker, C. Gouguenheim, I. Robinson, C. Sayers, and G. Gualdi. RFID and Sensing in the Supply Chain: Challenges and Opportunities. Technical Report HPL-2005-16, HP Labs, Palo Alto, CA, February 2005.

[49] J. Rao, S. Doraiswamy, H. Thakkar, and L. Colby. A Deferred Cleansing Approach for RFID Data Analytics. In *Proc. of the 32nd International Conference on Very Large Data Bases (VLDB'06)*, Seoul, Korea, September 2006.

[50] M.R. Rieback, B. Crispo, and A.S. Tanenbaum. Is Your Cat Infected with a Computer Virus? In *Proc. of 4th IEEE International Conference on Pervasive Computing and Communications (PerCom'06)*, Pisa, Italy, March 2006a.

[51] M.R. Rieback, B. Crispo, and A.S. Tanenbaum. The evolution of RFID security. *IEEE Pervasive Computing*, 7(1):62–69, January/March 2006b.

[52] M. Roberti. "DOD Releases Final RFID Policy." http://www.rfidjournal.com/article/articleview/1080/1/14/, August 2004.

[53] SAP. "SAP Auto-ID Infrastructure." http://www.sap.com/solutions/netweaver/autoidinfrastructure.epx, visited on 04/07/2006.

[54] SAP. "SAP NetWeaver." http://www.sap.com/solutions/netweaver/index.epx.

[55] S. Sarma. Integrating RFID. ACM Queue, 2(7):50–57, October 2004.

[56] E.W. Schuster, S.J. Allen, and D.L. Brock. *Global RFID: The Value of the EPCglobal Network for Supply Chain Management.* Springer, Heidelberg, Germany, 2007.

[57] R.C. Seacord, D. Plakosh, and G.A. Lewis. *Modernizing Legacy Systems: Software Technologies, Engineering Processes, and Business Practices.* Addison-Wesley, Reading, MA, 2003.

[58] J.R. Smith, K.P. Fishkin, B. Jiang, A. Mamishev, M. Philipose, A.D. Rea, S. Roy, and K. Sundara-Rajan. RFID-based techniques for human-activity detection. *Communications of the ACM*, 48(9):39–44, 2005.

[59] V. Stanford. Pervasive computing goes the last hundred feet with RFID systems. *IEEE Pervasive Computing*, 2(2):9–14, April/May 2003.

[60] Sun Microsystems. "Sun Java Enterprise System." http://www.sun.com/software/javaenterprisesystem/, visited on 10/09/2006.

[61] Sun Microsystems. "Sun Java System RFID Software." http://www.sun.com/software/ products/rfid/, visited on 10/09/2006.

[62] Sybase. "Sybase RFID Solutions." http://www.sybase.com/rfid, visited on 10/09/2006.

[63] G. Tolle, J. Polastre, R. Szewczyk, D. Culler, N. Turner, K. Tu, S. Burgess, T. Dawson, P. Buonadonna, D. Gay, and W. Hong. A Macroscope in the Redwoods. In *Proc. of the 3rd ACM Conference on Embedded Networked Sensor Systems (SenSys'05)*, San Diego, CA, November 2005.

[64] Venture Development Corporation (VDC). http://www.vdc-corp.com/, visited on 05/06/2006.

[65] F. Wang and P. Liu. Temporal Management of RFID Data. In *Proc. of the 31st International Conference on Very Large Data Bases (VLDB'05)*, Trondheim, Norway, September 2005.

[66] F. Wang, S. Liu, P. Liu, and Y. Bai. Bridging Physical and Virtual Worlds: Compex Event Processing for RFID Data Streams. In *Proc. of the 10th International Conference on Extending Database Technology (EDBT'2006)*, Munich, Germany, March 2006.

[67] R. Want. An introduction to RFID technology. *IEEE Pervasive Computing*, 7(1):25–33, January/March 2006.

[68] R. Want, K.P. Fishkin, A. Gujar, and B.L. Harrison. Bridging Physical and Virtual Worlds with Electronic Tags. In *Proc. of the SIGCHI Conference on Human Factors in Computing Systems (CHI'99)*, Pittsburgh, PA, May 1999.

[69] R. Weinstein. RFID: A technical overview and its application to the enterprise. *IEEE IT Professional*, 7(3):27–33, 2005.

[70] S.A. Weis, S.E. Sarma, R.L. Rivest, and D.W. Engels. Security and Privacy Aspects of Low-Cost Radio Frequency Identification Systems. In *Proc. of International Conference on Security in Pervasive Computing (SPC'03)*, Boppard, Germany, March 2003.

[71] J. Westhues. "Demo: Cloning a Verichip." http://cq.cx/verichip.pl, visited on 17/07/2006.

[72] R. Yu. "Electronic Passports Set to Thwart Forgers." *USA Today*, http://www.usa today.com/tech/news/2005-08-08-electronic-passports_x.htm, August 2005.

Chapter 5

RFID Security:
Threats and Solutions

Nicolas Sklavos and Vishal Agarwal

Contents

Low-cost Radio Frequency Identification (RFID) is a method of remotely storing and receiving data using devices called RFID tags. As the potential use of RFID has grown from tagging simple objects to more sophisticated uses, the RFID managers rely more on the reliability and integrity of the system. And thus they are continuously faced with the onslaught of new threats and vulnerabilities that puts critical assets, business operations, organization reputations, and personal security at risk. So the need for security has emerged as a key issue. In this chapter, we first provide a brief overview of RFID, the most recent standards related to this technology. Then we address both the risks and threats associated with the RFID security and finally we analyze the solutions proposed to date and their effectiveness.

5.1 Introduction

As its name implies, the term RFID is generally used to describe any technology that uses radio signals to identify specific objects. In practice, this means any technology that transmits specific identifying numbers using radio frequency. RFID technology is quickly evolving with improved efficiency in the supply chain, logistics, and inventory control. It enhances visibility of the movement of supplies providing opportunities for increased efficiency and also improves timeliness of information about goods. RFID tags can uniquely encode the individual identity of a particular product. Because many tags can be read at a distance (often measured in feet) by readers at known locations, they also provide information on location at time of read, and this information can be used to track tagged items. Consumer product manufacturers, suppliers, and retailers stand to benefit from RFID by knowing where goods are within and between businesses in the supply chain. The long-term goal of these organizations is to integrate RFID on the consumer level; However, without proper care, use of RFID could lead to many security concerns. EPCglobal Inc. [1] is a global standards organization commercializing the Electronic Product Code (EPC) and RFID worldwide. The main form of a barcode-type RFID device is known as an EPC tag. EPCglobal is walking with Universal Product Code (UPC) and European Article Number (EAN), the standards that regulate barcode use in the United States and the rest of the world, respectively. The vision of EPCglobal is a standardized system running on different platforms with a standardized protocol. It builds on existing technologies, such as servers, clients, databases, wireless communication, and Internet protocols, all with their own potential vulnerabilities (out of scope for the following discussion). While barcodes have historically been the primary means of tracking products, RFID systems are rapidly becoming the preferred technology for keeping tabs on people, pets, products, and even vehicles. One reason for this is because the read/write capability of an active RFID system enables the use of interactive applications. Also, the tags can be read from a distance and through a variety of substances, such as snow, fog, ice, or paint, where barcodes have proven useless (Figure 5.1).

RFID system is no contact, nonline-of-sight, and invisible identification, which is different from ubiquitous barcode identification system. Hence, it is difficult to completely stop the signals from being emitted from the tags. Tags are placed on pallets, cases, and individual items and can be scanned inches to meters away,

Figure 5.1 RFID tag with printed barcode.

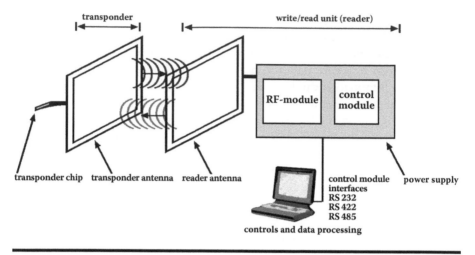

Figure 5.2   Layout and basic functions of a RFID system.

revealing the EPC number. The EPC number is the key to a database entry that contains information about the product and its owner. This has the potential to reduce purchase anonymity, and security advocates are worried about disclosing such information. RFID systems are composed of three key components: RFID tag (transponder), RFID tag reader (transceiver), and backend database. Figure 5.2 provides the basic layout of a RFID system.

5.1.1   RFID Tag (or Transponder)

Every object to be identified in a RFID system is physically labeled with a tag. Tags are typically composed of a microchip for storage and computation, and a coupling element, such as an antenna coil for communication.

RFID tags can be divided into two main categories:

1. Passive tags, which have no onboard power source. These tags receive their operational power from RFID reader devices. The tags antenna captures the radio frequency energy from the reader, stores it in a capacitor, and then uses it to power the tags logic circuits. After completing the requested commands, the tag uses the capacitors remaining energy to reflect a signal to the reader on a different frequency.

2. Active tags have onboard power sources, such as batteries. Active tags can support more sophisticated electronics with increased data storage, sensor interfaces, and specialized functions. In addition, they use batteries to transmit signals back to the RFID reader. This application is used in tracking cargo and collecting tolls electronically. They have read ranges of 100 m or more.

5.1.2 RFID Tag Reader (or Transceiver)

Tag readers interrogate tags for their data through an radio frequency (RF) interface. To provide additional functionality, readers may contain internal storage, processing power, or connections to backend databases. Computations, such as cryptographic calculations, may be carried out by the reader on behalf of a tag. If RFID tags become ubiquitous in consumer items, tag reading may become a desirable feature on consumer electronics.

5.1.3 Backend Database

Readers may use tag contents as a lookup key into a backend database, which can associate product information, tracking logs, key management information with a particular tag. Independent databases may be built by anyone with access to tag contents. This allows users along the supply chain to build their own applications. In many ways, tags are only useful if corroborated with a database in some way.

5.2 RFID Applications

RFID is here to stay. In the tracking inventory in a warehouse or maintaining a fleet of vehicles, there is a clear need for a fully automated data capture and analysis system that will help one keep track of his valuable assets and equipment. RFID provides unique solutions to difficult logistical tracking of inventory or equipment. It is particularly useful in applications where optically based systems fail and when read/write capabilities are required. The technology is stable and evolving, with open architectures becoming increasingly available.

The advantages of RFID over Barcode are:

- No line-of-sight requirement
- Long read range
- The tag can stand a harsh environment
- Multiple tags read/write
- Portable database
- Tracking people, items, and equipment in real-time

RFID finds a lot of application in various public and private sector fields. One such sector that is the focus of extensive RFID research is healthcare. In this field, RFID devices can be used to track equipment and people within a medical facility. Using RFID in different ways includes exploring how RFID can enhance the quality of elder care. By tagging key objects in a senior citizen home, such as prescription drug bottles, food items, and appliances, and embedding small RFID readers in gloves that can be worn by individuals, patients' daily habits can be monitored remotely by a caregiver.

RFID technology is a typical cross-section technology whose potential application can be found in practically all areas of daily life and business. Theoretically, the application areas of RFID systems are unlimited. From a cross-industry viewpoint, the following areas of applications can be applicable:

- Identification of objects
- Authentication of documents
- Theft protection
- Maintenance and repair
- Routing control
- Access authorization
- Environmental monitoring
- Supply chain management
- Automated payment
- Smart homes and offices

Much of the available market data on the use of RFID systems are limited to individual economic sectors and fail to give a comprehensive market overview. In the basic data used by various consulting firms, the survey methods and market classifications are very different from one another, not always understandable, and cannot be compared with one another. As a result, the status of diffusion, sales, and market shares of RFID systems remain unclear. The answer to the question as to whether RFID systems will be used in the future as mass technology depends on such factors as the success of the pilot projects running now.

With RFID, you can gain virtual real-time tracking capabilities of people, production, and manufacturing assets, such as monitor physical security of employees, manage access to secure, potentially hazardous work environments; and know where employees, contractors, and visitors are so you can expedite aid in emergency situations.

RFID tags are useful for a large variety of applications. For example, RFID tags are implanted in all kinds of personal and consumer goods, such as passports, partially assembled cars, frozen dinners, ski-lift passes, clothing, and public transportation tickets. Implantable RFID tags for animals also allow concerned owners to label their pets and livestock. Verichip Corp. has created a slightly adapted implantable RFID chip, the size of a grain of rice, for use in humans as well. Since its introduction, the Verichip was approved by the U.S. Food and Drug Administration. This tiny chip is currently deployed in both commercial and medical systems.

5.3 Introduction to Security Risks and Threats

While corporate giants tout the merits of RFID technology, civil liberties advocates point out that the ability to track people, products, vehicles, and even currency would create an Orwellian world: A place where law enforcement officials and nosy retailers could read the contents of a handbag perhaps without a person's knowledge simply by installing RFID readers nearby. Such a fear is not unfounded. As people start to rely on RFID technology, it will become easy to infer information about their behavior and personal tastes by observing their use of the products with this technology.

To make matters worse, RFID transponders also are too computationally limited to support traditional security enhancing technologies. This lack of information regulation between RFID tags and RFID readers may lead to undesirable situations. One such situation is unauthorized data collection, where attackers gather illicit information by either actively issuing queries to tags or passively eavesdropping on existing tag-reader communications. Even the RFID industry itself is aware of the threat to security posed by the development and installation of tags in commonplace items. Ignorance regarding the actual capabilities of RFID and the potential risk to consumer privacy and security will lead to unfounded fears and resistance from consumers and security advocates. Thus, this will slow the adoption and further development of an effective and secure technology.

The growing RFID security concern is one of the most pressing concerns for today's RFID executives. A RFID executive knows the stakes are high and the threats that inhabit RFID infrastructure will decelerate his business.

A problem very much related to privacy is tracking or violations of privacy. This actually is possible because many times tags provide the same identifier, which will allow a third person to establish a connection between a tag and its owner. Apart from the problems mentioned above, there are other important threats to RFID security as well.

5.3.1 Spoofing Identity (or Cloning)

Spoofing occurs when an attacker successfully impersonates a legitimate tag as, for example, in a man-in-the-middle attack. In spoofing, someone with a suitably programmed portable reader might covertly read and record a data transmission from a tag that could contain the tag ID (TID). When this data transmission is retransmitted, it appears to be a valid tag. Many people consider that "cloning."

However, the significant difference here is the form factor. A portable reader in a backpack or briefcase is obviously not the same as a tag or card. So, again, what you have is a copy of the tag's transmission, not a clone of the tag.

5.3.2 Physical Attacks

In order to mount these attacks, it is necessary to manipulate tags physically, generally in a laboratory. Some examples of physical attacks are probe attacks, material removal through shaped charges, circuit disruption and radiation imprinting. RFID tags offer very little resilience against these attacks.

5.3.3 Data Tempering

Data tempering occurs when an attacker modifies, adds, deletes, or reorders data. Rouge/clone tags, rouge/unauthorized readers, and side-channel attacks (interception of reader data by an unauthorized device) all threaten data security.

5.3.4 Deactivation

This type of attack renders the transponder useless through the unauthorized application of delete commands or kill commands, or through physical destruction. Depending on the type of deactivation, the reader no longer can detect the identity of the tag, or it cannot even detect the presence of the tag in the reading range.

5.3.5 Detaching the Tag

A transponder is separated physically from the tagged item and may subsequently be associated with a different item, in the same way that price tags are "switched." RFID systems are completely dependent on the unambiguous identication of the tagged items by the transponders. Thus, this type of attack poses a fundamental security problem, even though it may appear trivial at the first sight.

5.3.6 Eavesdropping

In these types of attacks, unintended recipients are able to intercept and read messages. The communication between reader and transponder via the air interface is monitored by intercepting and decoding the radio signals. This is one of the most specific threats to RFID systems.

5.3.7 Denial of Service (DoS)

Denial-of-service denies service to valid users. Denial-of-service attacks are easy to accomplish and difficult to guard against. A common example of this type of attack in RFID systems is the signal jamming of RF channels.

5.3.8 Falsification of Identity

In a secure RFID system, the reader must prove its authorization to the tag. If an attacker wants to read the data with his own reader, it must fake the identity of an authorized reader. Depending on the security measures in place, such an attack can be "very easy" to "practically impossible" to carry out. The reader might need access to the backend in order, for example, to retrieve keys that are stored there.

The attacker obtains the ID and any security information of a tag and uses these to deceive a reader into accepting the identity of this particular tag. This method of attack can be carried out using a device that is capable of emulating any kind of tag or by producing a new tag as a duplicate of the old one (cloning). This kind of attack results in several transponders with the same identity being in circulation.

5.3.9 Falsification of Contents

Data can be falsified by unauthorized write access to the tag. This type of attack is suitable for targeted deception, if the ID (serial number) and any other security information that might exist (e.g., keys) remains unchanged when the attack is carried out. This way the reader continues to recognize the identity of the transponders correctly. This kind of attack is possible only in the case of RFID systems that, in addition to ID and security information, store other information on the tag. (Figure 5.3)

5.4 Security Measures

Fortunately, as threats become faster and more aggressive, so does the response. In this section, we enumerate the various proposed effective approaches to the RFID security threat and risks. Also some new methods to the problem are highlighted.

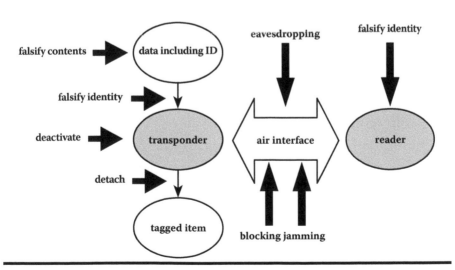

Figure 5.3 Basic types of attack in RFID systems.

5.4.1 Tag Killing Command or Permanent Deactivation

A kill command enables the anonymous nature of transponders by making the read-out of tags permanently impossible. This protects persons carrying tagged items from being surreptitiously identified and, thus, from being tracked.

To prevent wanton deactivation of tags, this kill command is pin protected. To kill a tag, a reader must also transmit a tag-specific PIN (32 bits long in the EPC Class-1 Gen-2 standard). Killing or discarding tags effectively enforces consumer privacy, but it eliminates all of the post-purchase benefits of RFID for the customer.

5.4.2 A Faraday Cage or Jamming Approach

In this approach, the RFID-labeled objects can be made secure by isolating them from any kind of electromagnetic waves. Faraday Cage is a metal or foil-lined container that is impenetrable to radio frequency waves. A RFID tag in a Faraday cage is effectively unreadable. It is also possible to jam the reading of RFID tags with devices that broadcast powerful, disruptive radio signals. Such jamming devices, however, in most cases, will violate government regulations on radio emissions.

5.4.3 Use of Blocker Tags

This is a privacy-protecting scheme proposed by Juels, Rivest, and Szydlo [2]. Their scheme depends on the incorporation into tags of a modifiable bit called a privacy bit. A 0 privacy bit marks a tag as subject to unrestricted public scanning and a 1 bit marks a tag as private. They refer to the space of identifiers with leading 1 bits as a privacy zone. A blocker tag is a special RFID tag that prevents unwanted scanning of tags mapped into the privacy zone. A blocker tag can be manufactured almost as cheaply as an ordinary tag. Blocking, moreover, may be adapted for use with ALOHA singulation protocol (the more common type).

5.4.4 Encryption

Even if the identifier emitted by a tag has no intrinsic meaning, it can still enable tracking; therefore, merely encrypting a tag identifier does not solve the problem. Thus, measures are headed that force tag identifiers to get suppressed or to change over time.

5.4.5 Rewriting

Ohkubo et al. [3] proposed an anonymous ID scheme. The fundamental idea of their proposal is to store anonymous ID, E (ID) of each tag, so that an adversary cannot know the real ID of the tag. E may represent a public or asymmetric key encryption algorithm, or a random value linked to the tag ID. In order to solve the tracking problem, the anonymous ID stored in the tag must be renewed by re-encryption as frequently as possible.

5.4.6 Minimalist Cryptography or Pseudonymization

While high-powered devices like readers can relabel tags for privacy, tags can alternatively relabel themselves. Juels [4] proposes a minimalist system in which every tag contains a small collection of pseudonyms; it rotates these pseudonyms, releasing a different one on each reader query. An authorized reader can store the full pseudonym set for a tag in advance, and therefore, identify the tag consistently. An unauthorized reader, one without knowledge of the full pseudonym set for a tag, is unable, therefore, to correlate different appearances of the same tag. Pseudonymization can hide the identity of the tag so that only authorized reader can find out its true identity.

5.4.7 Re-encryption

RFID labels will scroll out their EPC when requested. This unique identity carried by the RFID labels poses a security threat and a privacy threat. Thus, it is important to control access to a label's EPC, or not to allow a RFID label to respond with a nonidentifying response. Re-encryption offers a novel perspective on achieving these goals. Instead of storing the EPC in a write once format, it is possible, for instance, for a retailer in the supply chain to store an encrypted version of that EPC concatenated with a random number on the tag. This encryption may be performed using a secret key that is known solely to the retailer. Thus, when a label is requested to scroll out its EPC, it will scroll out an encrypted version, which to a third party will appear as a stream of random bits. In case of universal re-encryption the novelty in Golle et al.'s [5] proposal is that re-encryption neither requires nor yields knowledge of the corresponding public key under which a cipher text was computed.

5.4.8 Solutions Based on Hash Functions

Active querying attacks may be addressed by limiting who is permitted to read tag data through access control. Eavesdroppers may be dealt with by ensuring that tag contents are not broadcast in the clear over the forward channel. In the following section, we examine some of the security solutions proposed so far, based on Hash functions.

5.4.8.1 Hash Lock

The following scheme is proposed by Weis et al. [6] based on one-way hash functions. Each tag has a portion of memory reserved to store a temporary Meta ID

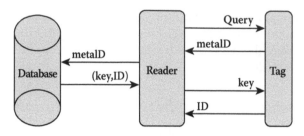

Figure 5.4 A reader unlocks a hash-locked tag.

and operates in either a locked or an unlocked state. The reader hashes a key k for each tag, and each tag holds a Meta ID (Meta ID = hash (k)). While locked, a tag answers all queries with its Meta ID and offers no other functionality. To unlock a tag, the owner queries the backend database with the Meta ID from the tag, looks up the appropriate key, and sends the key to the tag. The tag hashes the key and compares it to the stored Meta ID. (Figure 5.4)

5.4.8.2 Randomized Hash Lock

This procedure, proposed by Weis et al. [6], is based on the dynamic generation of a new Meta ID every time a read-out event occurs. Thus, the problem faced by a previous scheme that allows tracking is now avoided. To unlock a tag, the owner queries the backend database with the Meta ID from the tag, looks up the appropriate key, and sends the key to the tag. The tag hashes the key and compares it to the stored Meta ID (Figure 5.5).

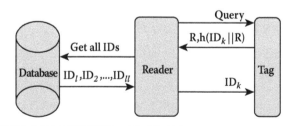

Figure 5.5 A reader unlocks a tag with randomized hash-lock scheme ID.

5.4.8.3   Hash Chain Scheme

Ohkubo et al. [3] suggest the chained hash procedure as a cryptographically robust alternative. At each activation, the tag calculates a new Meta ID, using two different hash functions. First, the current Meta ID is hashed in order to generate a new Meta ID, which is then hashed again with the aid of the second function. It is this second Meta ID that is transmitted to the reader. For the purpose of decoding, the reader must hash until a match with the Meta ID transmitted from the tag has been found. The advantage of this procedure is that it is not sensitive to repeated attempts to spy out the Meta ID during transmission via the air interface. An attacker would not be able to back calculate the Meta IDs that have been spied out. And, thus, it results that the anonymity of all preceding database entries (log entries) of the tag in question are preserved.

5.4.8.4   Tree-Based Algorithm

One of the main drawbacks of the hash schemes already proposed is that the load of the server (for identifying tags) is proportional to the number of tags. Molnar and Wagner [7] proposed a scheme in which a tag contains not one symmetric key, but multiple keys. This new protocol reduces the load to $O(\log n)$. This scheme results in striking improvement in efficiency, but there is a price to be paid for this efficiency gain. The tree structure creates an overlap among the sets of keys in tags. Compromise of the secrets in one tag, therefore, results in compromise of secrets in other tags. Compromise of a fraction of the tags in the system can lead to substantive privacy infringements. An offline delegation has been proposed by Molnar, Soppera, and Wagner [8].

Another approach to avoiding brute-force key search is for a reader to maintain a synchronized state with tags. Ohkubo et al. [3] and Juels [9] proposed some interesting approaches based on synchronization. Another interesting proposal is the work of Avoine and Oechslin [10], where a time-space trade-off is proposed.

5.4.9   Scheme by Henrici and Muller

Henrici and Muller [11] propose a procedure that makes possible the mutual authentication of tag and reader, as well as encryption of communication. This procedure also ensures the protection of "location privacy." In addition, no keys or other usable data are stored for any length of time on a tag, thus making physical attacks on the chip hardware uninteresting. The procedure gets by with a minimum exchange of information and is also resistant to interference on the transmission channel (air interface).

5.4.10 RFID Bill of Rights

It may prove valuable also to address RFID privacy and security issues through policy and regulation. In general, policy-based solutions are hard to implement and change, but have the advantage of being based on behavior and intent. Garfinkel [12] has proposed a RFID Bill of Rights that adapts the principles of fair information practices to RFID systems deployment.

5.4.11 RFID Guardian or Firewall

Rieback et al. [13] proposed a technique called RFID Guardian. The RFID Guardian is a mobile battery-powered device that offers personal RFID security and privacy management for people. The RFID Guardian monitors and regulates RFID usage, on the behalf of consumers. A Guardian (to use the first term) acts as a kind of personal RFID firewall. It intermediates reader requests to tags; viewed another way, the Guardian selectively simulates tags under its control.

5.4.12 Digital Signature-Based Scheme

Certicom Security for RFID Product Authentication [14] uses standards-based and proven cryptographic protocols for its RFID appliance, including a standardized public-key cryptography scheme from IEEE. This efficient ECC-based digital signature scheme enables a high level of security to be added to a RFID tag or reader without requiring a lot of computing power and storage. In fact, a 160-bit Elliptic Curve Cryptography (ECC) key provides the same level of protection as a 1024-bit RSA [9] key, but its digital signature is approximately one fourth the size. Also, when it is used to provide security for ECP, ECC saves two thirds the space compared to RSA at comparable strength. An ECC-based RFID solution is far more efficient than the methods developed so far using similar techniques. These solutions reduce bit storage, power consumption, and RAM required, thereby significantly improving operation speeds and overall performance.

5.4.13 Layered RFID Security Solution

Vandia [15] proposed and pioneered the layered RFID security model, which requires extending layers of security beyond the traditional IT network concept to encompass both physical and logical RFID infrastructure. The four layers approach is the first in its class to adequately ensure that the RFID infrastructure and its stakeholders are protected from all known threats and interconnect all the four layers with strong security links.

5.4.14 Detection Units and Screaming Tags

A RFID-enabled environment may be equipped with devices that can detect unauthorized reads or anomalous transmissions on tag operating frequencies. Detection units can be extended to detect when tags are disabled. Sarma [16] proposed the design of the tags that "scream" when killed. The schemes presented aid in detecting denial-of-service attacks.

5.4.15 Some other New Approaches

Tag readers have an asymmetric nature in its forward and backward channel when it comes to transmission of keys. Now if a tag reader needs to transmit some signal x to the tag via the forward channel, then that tag can generate a random signal y and transmit to the reader via the backward channel. Then the reader now sends the net signal, i.e., $x + y$ instead of only x due to the asymmetric nature. And, thus, eavesdroppers would not obtain any information about the original signal x.

Another way by which forward channel eavesdropping can be prevented is to broadcast 'chaff' commands from the reader, intended to confuse or dilute information collected by eavesdroppers. By negotiating a shared secret, these commands can be filtered, or 'winnowed' by tags using a simple media access control (MAC).

To enable end users to access the functionality of tags affixed to items they have purchased, a master key can be printed within a product's packaging as a barcode or decimal number.

5.4.16 Tag/Reader Deactivation Using Induced Fields

RFID tags or readers can be deactivated using strong electromagnetic fields with the help of some specific burnout points or canceling the signal transfer using electromagnetic radiation. Not much work is being done in this direction so far.

5.4.17 Direction Sensitive Tag Entry

RFID tags read the same 32 bit tag entry, irrespective of the direction the tag is read. Now, if a RFID system is designed in such a way that a tag's actual value would get read only if it is inclined at some specific angle, say theta, then the tag's data information would be prevented from any unauthorized reader. To read the tag, that reader would have to orient itself in such a way so as to get the angle, theta. Thus, the chances of theft to security is minimized. In Figure 5.6, a different theta refers to a different tag value.

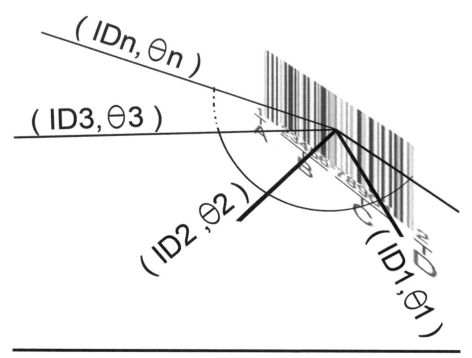

Figure 5.6   Direction-sensitive tag value.

5.5 Challenges

The economic success of RFID technologies will depend not only on technical possibilities, but in addition to technology and standardization, the market and price developments, the requirements on information security, and data protection have to be considered along with social discourse in the context of RFID. The primary challenge in providing security and access control mechanisms in low-cost RFID is scarcity of resources. As mentioned above, tags will only have a fraction of the gate count available in smart cards. Security mechanisms of passively powered tags will need to be carefully designed so as not to leave tags in an insecure state in the event of power loss or interruption. Flexibility and openness of design are of utmost importance to a successful RFID system. Future tag developments will allow greater storage, faster performance, and new functionality to be incorporated into low-cost tag designs. Current security mechanisms for RFID systems should not impede utilization of future technologies, nor should they adversely affect the user experience. Retailers will not adopt RFID systems if they necessarily hinder the consumer check-out process. No one expects customers to undergo complicated security procedures every time they purchase a quart of milk. On the other hand, RFID tags should be as open a platform as possible, supporting both existing applications and applications yet to be conceived. Security features should not interfere with the development of new applications by third parties. Many useful consumer applications could emerge from grass-roots development and should not be impaired by proprietary or closed security mechanisms.

5.6 Conclusions and Outlook

As the RFID market expands, users will see the continued proliferation of RFID tags built for highly specialized vertical markets, which means greater variety and the consequent need to ensure interoperability. A great deal of research and development is currently under way in the RFID security field to mitigate both known and postulated risks. Manufacturers, business managers, and RFID systems engineers continue to weigh the trade-offs between chip size, cost, functionality, security, and privacy with the bottom-line impact on business processes. In the coming months, security features supporting data confidentiality, tag-to-reader authentication, optimized RF protocols, high-assurance readers, and secure system engineering principles should become available. Security and privacy in RFID tags are not just technical issues, important policy questions arise as RFID tags join to create large sensor networks and bring us closer to ubiquitous computing. With public attention focused on the RFID landscape, security and privacy have moved to the forefront in RFID standards work and the results will be worth watching.

References

[1] *EPCTM Radio-Frequency Identity Protocols Class-1 Generation-2 UHF RFID Protocol for Communications at 860 MHz – 960 MHz, ver. 1.0.9.* EPCglobal Inc., Jan. 31, 2005. Available: http://www.epcglobalinc.org/.

[2] A. Juels, R.L. Rivest, and M. Szydlo. *The blocker tag: Selective blocking of RFID tags for consumer privacy.* In V. Atluri, ed. 8th ACM Conference on Computer and Communications Security (Philadelphia, PA, November, 2001) pp. 103–111. ACM Press, 2003.

[3] M. Ohkubo, K. Suzuki, and S. Kinoshita. *Efficient hash-chain based RFID privacy protection scheme.* In International Conference on Ubiquitous Computing–Ubicomp, Workshop Privacy: Current Status and Future Directions, 2004.

[4] A. Juels. *Minimalist Cryptography for RFID Tags.* 4th Conference on Security in Comm. Networks (SCN), C. Blundo and S. Cimato, eds., Springer-Verlag, Heidelberg, Germany, 2004, 149–164.

[5] P. Golle, M. Jakobsson, A. Juels, and P. Syverson. *Universal re-encryption for mixnets.* In T. Okamoto, ed., RSA Conference Cryptographers Track (CT-RSA), vol. 2964 of Lecture Notes in Computer Science, pp. 163–178, 2004.

[6] S. Weis, S. Sarma, R. Rivest, and D. Engels. *Security and privacy aspects of low-cost radio frequency identification systems.* In D. Hutter, G. Muller, W. Stephan, and M. Ullmann, eds. International Conference on Security in Pervasive Computing SPC 2003, vol. 2802 of Lecture Notes in Computer Science, pp. 454–469. Springer-Verlag, Heidelberg, Germany, 2003.

[7] D. Molnar and D. Wagner. *Privacy and security in library RFID: Issues, practices, and architectures.* In B. Pflitzmann and P. McDaniel, eds., ACM Conference on Communications and Computer Security , pp. 210–219. ACM Press, 2004.

[8] D. Molnar, A. Soppera, and D. Wagner. *A scalable, delegatable pseudonym protocol enabling ownership transfer of RFID tags.* In B. Preneel and S. Tavares, eds. Selected Areas in Cryptography, SAC 2005, Lecture Notes in Computer Science. Springer-Verlag, Heidelberg, Germany, 2005.

[9] A. Juels. *Minimalist cryptography for low-cost RFID tags.* In C. Blundo and S. Cimato, eds. The Fourth International Conference on Security in Communication Networks SCN 2004, vol. 3352 of Lecture Notes in Computer Science, pp. 149–164. Springer-Verlag, Heidelberg, Germany, 2004.

[10] G. Avoine and P. Oechslin. *A scalable and provably secure hash-based RFID protocol.* In F. Stajano and R. Thomas, eds. The 2nd IEEE International Workshop on Pervasive Computing and Communication Security Parsec 2005, pp. 110–114. IEEE Computer Society Press, Washington, D.C., 2005.

[11] D. Henrici and P. Muller. *Tackling Security and Privacy Issues in Radio Frequency Identification Devices.* In Ferscha A., Mattern F.: Pervasive Computing (Proceedings of PERVASIVE 2004, 2nd International Conference on Pervasive Computing) pp. 219–224. Springer-Verlag, Heidelberg, Germany, 2004.

[12] S. Garfinkel. *An RFID Bill of Rights.* Technology Review, p. 35, October 2002.

[13] M. Rieback, B. Crispo, and A. Tanenbaum. *RFID Guardian: A battery-powered mobile device for RFID privacy management.* In C. Boyd and J.M. Gonzalez Nieto, eds. Australian Conference on Information Security and Privacy – ACISP 2005, vol. 3574 of Lecture Notes in Computer Science, pp. 184–194. Springer-Verlag, Heidelberg, Germany, 2005.

[14] *Certicom Security for RFID Product Authentication.* http://www.certicom.com/index.php?action=sol,rfid.

[15] *Vandia Security Solutions.* http://www.vandia.com/, referenced 2007.

[16] S.E. Sarma. *Personal correspondence*, 2002.

Further Reading

[1] R.J. Anderson and M.G. Kuhn. *Low cost attacks on tamper resistant devices.* In B. Christianson, B. Crispo, T.M.A. Lomas, and M. Roe, Eds., Security Protocols Workshop, vol. 1361 of Lecture Notes in Computer Science, 125–136. Springer-Verlag, Heidelberg, Germany, 1997.

[2] G. Avoine, E. Dysli, and P. Oechslin. *Reducing time complexity in RFID systems.*

[3] S. Bono, M. Green, A. Stubblefield, A. Juels, A. Rubin, and M. Szydlo. *Security analysis of a cryptographically-enabled RFID device.* In Proc. USENIX Security Symposium, July–August 2005.

[4] T. Dimitriou. *A lightweight RFID protocol to protect against traceability and cloning attacks.* In Proc. of SecureComm'05, Athens, Greece, August–September, 2005.

[5] K. Finkenzeller. *RFID Handbook.* John Wiley & Sons, New York, 1999.

[6] K. Fishkin and J. Lundell. *RFID in healthcare*. In S. Garfinkel and B. Rosenberg, eds. RFID: Applications Security, and Privacy, pp. 211–228. Addison-Wesley, Reading, MA, 2005.

[7] S. Garfinkel and B. Rosenberg, eds. *RFID Applications, Security, and Privacy*. Addison-Wesley, Reading, MA, 2005.

[8] A. Juels. *Strengthing EPC tags against cloning*. In ACM Workshop on Wireless Security (WiSe). ACM Press, 2005.

[9] Ari Juels. *RSA Security—RFID Security and Privacy: A Research Survey*.

[10] P. Peris-Lopez, J.C. Hernandez-Castro, J.M. Estevez-Tapiador, and A. Ribagorda. *An Efficient Mutual-Authentication Protocol for Low-cost RFID Tags*.

[11] *RFID Journal*. Online publication. Referenced 2005.

[12] *Texas Instruments gen-2 inlay data sheet*, 2005. Referenced 2005 at: http://www.ti.com/rfid/docs/manuals/.

[13] I. Vajda and L. Buttyian. *Lightweight authentication protocols for low-cost RFID tags*. In Proc. of UbiComp'03, Seattle, WA, 2003.

[14] S.A. Weis. *Radio-frequency identification security and privacy*. Masters thesis, M.I.T., June 2003.

Chapter 6

RFID Specification Revisited

Pedro Peris-Lopez, Julio C. Hernandez-Castro,
Juan M. Estevez-Tapiador, and Arturo Ribagorda

Contents

RFID (Radio Frequency Identification) is the name given to all technologies that use radio waves to automatically identify and account transactions on people, animals, or objects [1] by means of electromagnetic proximity [2]. RFID technology is not new, as one of its first usages dates from 1940 where a RFID-based Identification Friend or Foe (IFF) system was used [3]. There are multiple standards related to RFID technology. In this chapter, the Electronic Product Code (EPC) Class-1 Generation-2 is examined. This standard can be considered as the "universal" standard for Class-1 RFID tags. Class-1 RFID tags are very limited both in their computational and storage capabilities. Because of these severe restrictions, the usage of standard cryptographic primitives is not possible. However, RFID tags are susceptible to attacks also found in other technologies, such as wireless, bluetooth, smart-cards, etc. Therefore, once the EPC Class-1 Generation-2 specification is explained, a security analysis will reveal its weak points. Furthermore, current proposals to enhance its security level are presented and analyzed. Finally, the chapter is concluded identifying some open research issues to increment the security of low-cost RFID tags.

6.1 Introduction

RFID systems are already used for a large number of applications related to object identification: retail stock management, access control, animal tracking, theft prevention, sports timing, medical applications, etc. In fact, RFID applications are only limited by human imagination. However, a number of security questions are still open. As it has been shown by a recent European Union (EU)-funded public consultation [4], most of the participants are seriously concerned about privacy implications and tracking, or the violation of location privacy, which was the most important issue.

RFID is a relatively heterogeneous radio technology with a significant number of associated standards. As in Peris-Lopez et al. [5], standards can be classified according to five main categories: contactless integrated circuit cards, RFID in animals, item management, near field communication (NFC), and EPC. Figure 6.1 summarizes the most important of those.

The benefits of standards are clear and assumed by almost everyone. The growth of any new technology is in many cases due in part to the establishment of open standards. In 2003, there was a clear lack of harmonization and major RFID vendors offered mainly proprietary systems. Fortunately, things are quickly changing. Today, EPCglobal [6] and ISO [7] join forces to publicize and harmonize the use of RFID technology.

The EPCglobal is a joint venture between EAN International and the Uniform Code Council (UCC). EPCglobal fulfills the industry's need for an effective RFID network standard with its EPC (Electronic Product Code) Network. Within EPC-global, the Hardware Action Group (HAG) develops specifications for hardware components of the EPC Network, including tags and readers. The EPC system defines four RFID tag classes:

1. **Class-1: Identity Tags.** Passive backscatter tags with the following minimum features:
   - An Electronic Product Code (EPC) identifier
   - A tag identifier (TID)
   - A kill function that permanently disables the tag
   - Optional password-protected access control and optional user memory
2. **Class-2: Higher functionality tags.** Passive tags with all the aforementioned features and also including the following:
   - An extended TID
   - Extended memory
   - Authenticated access control
   - Additional features (TBD), as will be defined in the Class-2 specification
3. **Class-3:** Semipassive tags with all the aforementioned features and also including the following:
   - An integral power source
   - An integrated sensing circuitry
4. **Class-4:** Active tags with all the aforementioned features and also including the following:
   - Tag-to-tag communications
   - Active communications
   - Ad hoc and networking capabilities

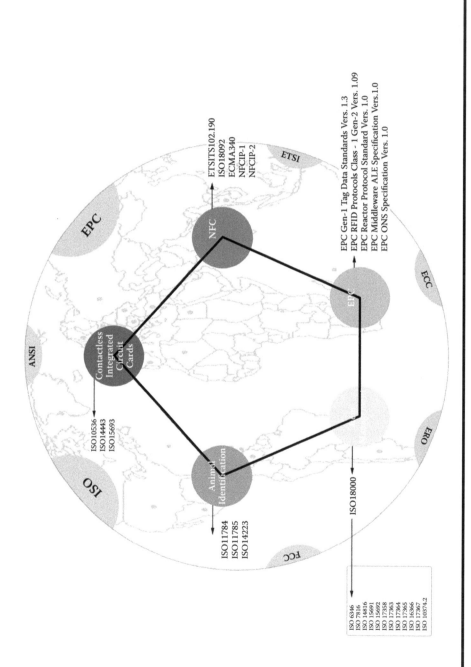

Figure 6.1 **RFID standards.**

6.2 Generation-2 Versus Generation-1

One of the most important standards proposed by EPCglobal is the EPCglobal Class-1 Gen-2 RFID specification [8]. This standard was adopted by EPCglobal in 2004 and was sent to ISO. Eighteen months later (March to April 2006), it was ratified by ISO and published as an amendment to its 18000-6 standard.

These specifications provide a great advance in consolidating the adoption of RFID technology [9]. Where previously there were several specifications, such as EPC Class-1 and EPC Class-0, a single ultra high frequency (UHF) specification is now established. In order to provide a worldwide deployment, emerging UHF regulations in different regions have been taken into account. Additionally, the best features of the preceding specifications have been improved, and a range of future applications including higher function sensor tags have been foreseen.

6.2.1 Read and Write Speed

Generation-1 provides a single communication speed, providing satisfactory speed and adequate robustness for most applications. On the other hand, four different communication speeds are available in Gen-2 to provide more flexibility for different operational environments. Gen-2 tags have a maximum theoretical reading speed of around 1,000 tags per second (when insulated from RF noise), but in very noisy environments that speed is reduced to around 100 tags/sec. The read speed of Gen-2 is then about twice as fast as in Gen-1 in real conditions with average read rates of around 500 tags per second.

Furthermore, Gen-2 specifies the speed at which tags can be programmed. The specification dictates that tags should be writable at a minimum rate of about 5 per second, setting 30 tag/sec as the objective value in optimum conditions.

6.2.2 Robust Tag Counting

Tags compliant with Gen-1 specification are singulated by means of binary tree walking protocols with persistent sleep/wake states. The Gen-2 specification are based on the principle that tags experiencing brief moments of power can be read. Thusly, the *Q protocol*, which is based on simple query/acknowledgment exchanges between reader and tags, is employed. *Q* is a parameter that a reader uses to regulate the probability of tag response. Briefly, a reader sends a query to a tag and the tag loads a *Q*-bit random number (or pseudo-random number) into its slot counter. Tag responds with a random number when the value in its slot counter is fixed to zero. The reader then sends an acknowledgement that includes the tag's random number, which then sends back its ID (i.e., EPC). The process continues until all tags in the reader's field have been counted.

In Gen-1, tags are switched to sleep mode after they have been read, easing the reading of tags that have not yet been counted. To begin a new count, a wake-up message is sent to tags in order for them to be ready for reading. Multiple wake-up, count, and sleep cycles are necessary to ensure that all tags in a reader field have been read. Gen-2 refines this process by introducing a dual state, avoiding the necessity of a wake-up command. Under this approximation, tags change their state each time they are read. As the reader counts tags in "A" state, these change automatically to "B" state, and vice versa. Gen-2 repeats counts of A and B until all tags have been identified. Therefore, these two mechanisms allow both to increment the reading speed and to make sure all tags have been counted.

6.2.3 Dense Reader Operation

If there are many readers operating and querying in close proximity at the same time, this can drown out the weak responses of tags. In the United States, frequency hopping is used, as there are no severe bandwidth requirements in the UHF band. In Europe, the band available to RFID is relatively narrow, so this approach is not possible. Readers are required to "listen before talking." First they determine that the channel is not already in use; only then can they start communication.

Gen-2 tries to improve its features under dense reader operation in several ways. First, the available RF bandwidth is used as efficiently as possible, only minimal data is exchanged between readers and tags. Second, Gen-2 provides new radio signaling techniques, which easily allows isolatation of tag response, even in noisy conditions. Therefore, "Miller subcarrier" or "FM0" are employed. Third, three modes of operation are available: single reader, multireader and dense reader. Finally, Gen-2 verifies data as in Gen-1 specification to ensure accurate reads in noisy environments and adds a feature that confirms when tags have been correctly written.

6.2.4 Parallel Counting

It is possible that several readers can simultaneously communicate with the same tag. Under this situation, a tag might change its state in the middle of another reader's query, causing the loss of the tag by the second reader. To solve this problem, tag support "sessions" allowing a single tag to communicate with two or more readers. Four logical sessions are available to be assigned to readers.

6.3 EPC Class-1 Generation-2 Specification

The EPC Class-1 Generation-2 specification [8], in the following EPC-C1G2, defines the physical and logical requirements for RFID systems operating in the 860 to 960 MHz frequency range. These systems are made up of two main components: interrogators, also known as readers; and tags, also known as labels.

Modulating a RF signal (860 to 960 MHz), a reader transmits information to a tag. As tags are passive, all of their operating energy is received from the reader's RF waveform. Indeed, both information and operating energy are extracted from the signal sent by the reader.

Furthermore, as tags do not have a power source, tags can only answer after a message is sent by the reader. These kind of systems are known as Interrogator Talk First (ITF). A reader receives information from a tag by transmitting a continuos wave RF signal to the tag. The tag backscatters a signal to the reader by means of the modulation of the reflection coefficient of its antenna. Communications are half-duplex, so the reader talks and the tag listens, or vice versa.

6.3.1 Physical Layer

A reader sends information to one or more tags by modulating a RF carrier using double-sideband amplitude shift keying (DSB-ASK), single-sideband amplitude shift keying (SSB-ASK), or phase reversal amplitude shift keying (PS-ASK) using pulse interval encoding (PIE) format. Tags receive their operating power from this RF signal.

In order to receive information from a tag, readers transmit an unmodulated RF carrier and listen for a backscattered replay. Tags send information by backscatter, modulating the amplitude/phase of the RF carrier. The encoding format is either a FM0 or Miller-modulated subcarrier.

6.3.2 Tag Identification Layer

A reader interacts with tags using three basic operations:

1. **Select:** The operation of choosing a subset of the tag population for inventory and access. Working with databases, this operation is similar to selecting records.
2. **Inventory:** The operation of identifying tags. Concretely, after the exchange of several messages (inventory round), the tag sends to the reader the protocol control (PC), EPC, and a cyclic redundancy check (CRC)-16 value. An inventory round operates in one and only one session at a time.
3. **Access:** The operation of communicating (reading from and/or writing to) with a tag, which is comprised of multiples commands. Tags have to be unequivocally identified before access.

6.3.3 Tag Memory

Tag memory is separated logically into four banks, as illustrated in Figure 6.2.

1. **Reserved memory (Bank 00):** This area of memory shall contain the kill and access passwords. Unless this memory locations have nonzero values,

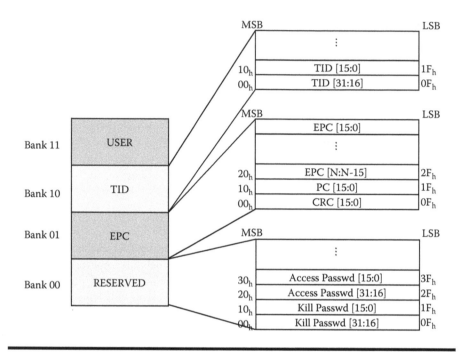

Figure 6.2 Logical memory map.

the kill and access commands will not be accepted. Furthermore, these locations cannot be locked or protected without invoking the access command.

The kill and access passwords are 32-bits values. Once the tag receives the kill password, it renders it silent thereafter. Tags with a nonzero access password, have to receive this before transitioning to a secure state.

2. **EPC memory (Bank 01):** This area of memory shall contain a CRC-16 checksum. Thus, the complement-one of the precursor defined in ISO/ IEC 13239 is computed. A basic integrity check is implemented by using a CRC-16 checksum of the PC and EPC values that a tag backscatters during an inventory operation.

    In the same block, we can find the PC bits and a code (such as an EPC) that unequivocally identifies the object to which the tag is attached. The PC is subdivided into the EPC length field, RFU (reserved for future use), and a number system identifier (NSI).

3. **TID memory (Bank 10):** This area of memory shall contain an 8-bit ISO/ IEC 15693 class identifier. Additionally, sufficient information to unequivocally identify the custom commands and/or optional features supported by the tag also are stored.

4. **User (Bank 11):** This area of memory allows user-specific data storage. The memory organization is user-defined.

6.3.4 Tag States and Slot Counter

As defined in EPC-C1G2, tags shall implement different states, as displayed in Figure 6.3 and Figure 6.4.

- **Ready state:** After being energized, a tag that is not killed shall enter a ready state. The tag shall remain in this ready state until it receives an accurate *Query* command. Tag loads a *Q*-bit number from its random number generator (RNG), and transitions to the arbitrate state if the number is nonzero, or to the replay state if the number is zero.
- **Arbitrate state:** A tag in an arbitrate state will decrease its slot counter every time it receives a *QueryRep*, transitioning to the replay state and backscattering a RN16 when its slot counter reaches 0000_h.
- **Replay state:** A tag shall backscatter a RN16 once entering in replay state. If the tag receives a valid acknowledge (ACK), it shall transition to "acknowledge" state, backscattering its PC, EPC, and CRC-16. Otherwise, the tag remains in "arbitrate" state.
- **Acknowledge state:** A tag in acknowledge state may transition to any state except "killed."
- **Open state:** After receiving a *Req_RN* command, a tag in "acknowledge" state whose access password is nonzero shall transition to "open" state. The tag backscatters a new RN16 that both reader and tag shall use in subsequence messages. Tags in an "open" state can execute all access commands except *Lock* and may transition to any state except "acknowledge".
- **Secured state:** A tag in "acknowledge" state whose access password is zero shall transition to "secured" state, upon receiving a *Req_RN* command. The tag backscatters a new RN16 that both reader and tag shall use in subsequence messages. A tag in the "open" state whose access password is nonzero shall transition to "secured" state, after receiving a valid access command, which includes the same *handle* that it previously backscattered when it transitioned from "acknowledge" state to the "open" state. Tags in

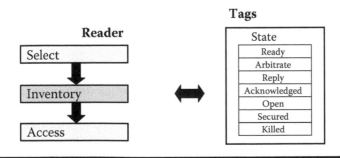

Figure 6.3   Tags state diagram.

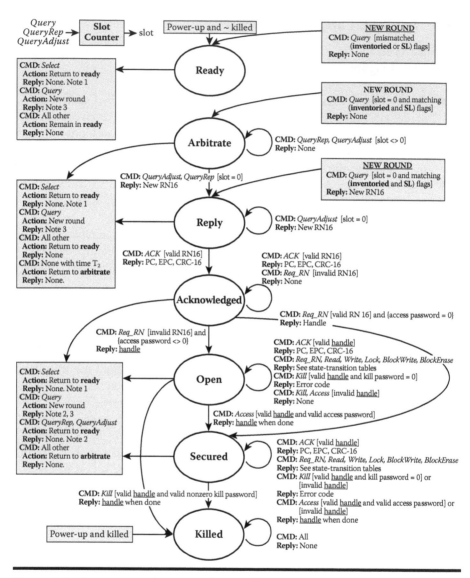

Figure 6.4   Interrogator/tag operations and tag state [8].

"secured" state can execute all access commands and may transition to any state except "open" or "acknowledge."

■ **Killed state:** Once a kill password is received by a tag in either "open" state or "secured" state, it shall enter the "killed" state. Kill permanently disables a tag. A tag shall notify the reader that the killed operation was successful and shall not respond to any reader thereafter.

Tags shall implement a 15-bit slot counter. Once a *Query* or *QueryAdjust* command is received, a tag shall load into its slot counter a value between 0 and $2^Q - 1$ obtained from a tag's pseudorandom number generator (PRNG). Q is a integer in the range (0, 15). A *Query* specifies Q, and a *QueryAdjust* may modify Q from the prior *Query*.

6.3.5 Managing Tag Populations

As shown in Figure 6.3, tag populations are managed using three basic operations. Each of these operations comprise one or more commands. The operations are defined as follows:

1. **Select:** The process that allows a reader to select a subset of the tag population for inventory and access. The select command is *Select*. A reader may use one or more *Select* commands to select a particular tag previous to inventory.
2. **Inventory:** The process that allows a reader to identify a tag. The inventory operation is started by transmitting a *Query* command in one of the four sessions that the tag can handle. One or more tags may replay. The reader isolates a single tag response and requests the PC, EPC, CRC-16 from the tag. An inventory round can only operate one session at a time. The inventory command set is comprised of the following commands: *Query*, *QueryAdjust*, *QueryRep*, *ACK*, *NACK*. All of these commands are mandatory.
3. **Access:** The process that allows a reader to interact (read from or write to) with individual tags. Tags have to be unequivocally identified prior to access. There are multiple access commands, some of which employ a one-time, pad-based cover coding in the $R \Rightarrow T$ link. The set of access commands is composed of the following mandatory commands: *Req_RN*, *Read*, *Write*, and *Lock*. Additionally, there are other optional commands described in the specification: *Access*, *Blockwrite*, and *BlockErase*.

6.4 Risks and Threats

One of the main problems that ubiquitous computing has to solve before it is widely developed is privacy [10]. Products labeled with insecure tags could reveal sensitive information when queried by readers. Readers frequently are not authenticated and tags usually answer in a complete transparent way. Moreover, even if we assume that tag's contents are secure, tracking (violation of location privacy) protection is not guaranteed. Tags usually answer with the same identifier. These predictable tag answers allow a third party to establish an association between tags and their owners. Even if tags only contain product codes, rather than a unique serial number, tracking can still be possible by using an assembly of tags (constellation) [11].

Although the two problems mentioned above are the most important security questions that arise in RFID technology, there are some others worth mentioning:

■ **Physical attacks:** In order to mount these attacks, it is necessary to manipulate tags physically, generally in a laboratory. Some examples of physical attacks are probe attacks, material removal through shaped charges or water etching, radiation imprinting, circuit disruption, and clock glitching, among others.

- **Denial of Service (DOS):** A common example of this type of attack in RFID systems is the signal jamming of RF channels.
- **Counterfeiting:** Attacks that consist in modifying the identity of an item by means of tag manipulation.
- **Spoofing:** An attacker is able to successfully impersonate another, for example, in a man-in-the-middle attack.
- **Eavesdropping:** Attacks when unintended recipients are capable of intercepting and reading messages.
- **Traffic analysis:** The process of intercepting and examining messages in order to extract information from patterns in communication. It can be performed even when the messages are encrypted and cannot be decrypted. In general, the greater the number of messages observed, the more information that can be inferred from the traffic.

6.5 Cryptographic Primitives

Traditional cryptographic primitives exceed the capabilities of low-cost RFID tags, such as those compliant with EPC-C1G2. The required hardware complexity of these devices may be weighted up by its circuit area or the number of equivalent logic gates. At most, around 4 K gates are assumed to be devoted to security-related tasks [12].

6.5.1 Hash Functions

The best implementation of SHA-256 requires around 11 K gates and 1,120 clock cycles to perform a hash calculation on a 512-bit data block [13]. As the number of needed resources are quite higher than those of a low-cost RFID tag, it may seem natural to propose the use of smaller hash functions. However, functions, such as SHA-1 (8.1 K gates, 1,228 clock cycles) or MD5 (8.4 K gates, 612 clock cycles), cannot be fitted in a low-cost tag [13]. Recently, some authors suggested the usage of a new "universal hash function" [14]. Although this solution only needs around 1.7 K gates, a deeper security analysis is needed and has not yet been accomplished. Furthermore, this function has only a 64-bit output, which does not guarantee an appropriate security level because finding collisions is a relatively easy task due to the birthday paradox* (around 2^{32} operations).

---

* The birthday paradox states that if there are 23 people in a room, then there is a slightly more than 50:50 chance that at least two of them will have the same birthday.

6.5.2 Ciphers

As traditional ciphers (DES, AES, etc.) exceed by large the computational and storage capabilities of constrained devices [15,16], some authors are focusing on efficient implementations of these ciphers. We can point out the interesting work of Feldhofer et al. who have proposed an efficient AES implementation (3,400 gates, 12 kbps) [16,17].

Another interesting work is the eSTREAM Project, which tries to identify new secure stream ciphers. The profile 2 is focused on stream ciphers for hardware applications with restricted resources, such as limited storage, gate count, or power consumption. Two of the candidates that have been selected for the Phase 3 (profile 3) are Grain and Trivium. Grain (2,133 gates, 100 kbps) and Trivium (3,000 gates, 100 kbps) are synchronous stream ciphers proposed by Hell et al. [18,19] and De Canniere et al. [18,20]. Despite the fact that these ciphers may fit in a low-cost RFID tag, the security of these ciphers is still an open question, as shown by the attacks [21,22] and the publication of amendments and different versions (Grain, Grain 1.0, Grain-128) [23].

6.5.3 Pseudorandom Number Generators

The need for random and pseudorandom numbers arises in many cryptographic applications. In fact, the usage of PRNGs in RFIDs has been proposed since their very beginning. In 2003, Weis et al. [11] proposed the randomized hash-locking scheme, based on a hash function and a random number generator, in order to prevent tracking, but limiting its applicability to small tag populations [24]. Molnar and Wagner [25] proposed a simple protocol for enhancing passwords in RFID tags. There are other papers where the use of a PRNG has been proposed [26–30]. Nowadays, the used of a PRNG has been ratified by EPCglobal (EPC-C1G2) and ISO (ISO/IEC 18006-C). On the contrary, tags compliant with EPC-C1G2 do not support onboard conventional encryption, hash functions, or any other kind of cryptographic primitive.

According to the EPC-C1G2, tags shall generate 16-bit random or pseudorandom numbers (RN16), and shall have the ability to extract Q-bit subsets from a RN16 to preload into its slot counter. Additionally, tags should be able to temporarily store at least two RN16s while powered, e.g., a handle and a 16-bit cover-code during password transactions. The generator (RN16) should meet the following randomness criteria:

- **Probability of a single RN16:** The probability that any RN16 drawn from the RNG has value RN16 = j, for any j shall be bounded by $0.8/2^{16} < P(RN16 = j) < 1.25/2^{16}$.
- **Probability of simultaneously identical sequences:** For a tag population of up to 10,000 tags, the probability that any of two or more tags

simultaneously generating the same sequence of RN16s shall be less than 0.1 percent, regardless of when the tags are energized.

■ **Probability of predicting an RN16:** An RN16 drawn from a tag's RNG 10 ms after the end of T_r, shall not be predictable with a probability greater than 0.025 percent if the outcomes of prior draws from RNG, performed under identical conditions are known.

6.6 Security Analysis and Open Issues

In this section, we see in detail the messages interchanged in the different operations between tags and readers. Once the procedures has been understood, the security faults can be easily identified.

6.6.1 Inventory Procedure

A *Query* initiates an inventory round and decides which tags participate in the round. After receiving the *Query*, tags shall pick up a random value in the range (0, $2^Q - 1$), and load this value into their slot counter. Tags that pick up a nonzero value shall transition to the arbitrate state and await a *QueryAdjust* or *QueryResp* command. Assuming that a single tag answers (slot_counter = 0), the query response algorithm proceeds as follows:

■ $T \Rightarrow R$: The tag backscatters an RN16 and enters in replay state.
■ $R \Rightarrow T$: The reader acknowledges the tag with an *ACK* containing the same RN16.
■ $T \Rightarrow R$: The acknowledged tag transitions to the acknowledged state, and backscatters its PC, EPC, and CRC-16.
■ $R \Rightarrow T$: The reader sends a *QueryAdjust* or *QueryRep*, so the identified tag inverts its inventory flag (i.e., A → B or B → A) and transitions to the ready state.

The security of this procedure is nonexistent. Imagine a scenario where a passive eavesdropper is listening to the channel. Under these conditions, the following security drawbacks are:

■ Tags do not transmit the EPC in a secure way. Instead, the EPC is transmitted in plain-text. So, one of the most important concerns about the use of RFID technology is not accomplished. In other words, the privacy information of the tag is easily jeopardized.

- Every time a tag is interrogated, it always transmits the same EPC. Due to the fact that the EPC transmitted by a tag is fixed, a tag may be associated with its holder allowing its tracking. Suppose that your watch has a tag compliant with the EPC-C1G2. An attacker may place readers in your favorite entertainment places: cinemas, pubs, shops, etc. During two months, the attacker's readers store the day and hours when you stay at these places. Then, the attacker collects all this information, obtaining a consumer profile of you, which is very valuable information to a great number of companies. It is only an example of how privacy location may be compromised. Therefore, another of the citizen concerns about the implantation of RFID technology is again questioned.

The above scenario shows how privacy and location privacy is not guaranteed even under a weak attack scenario where there is a passive eavesdropper. To avoid these two connected problems, researchers propose the use of pseudonyms. In general terms, a pseudonym is a fictitious name. In RFID context, a pseudonym is interpreted as an anonymized static identifier. However, under these conditions only privacy is guaranteed, as private information is passed on the channel. An additional requirement is needed to prevent an attacker from being able to track a holder's tag. Thusly, every time the tag is interrogated, the tag has to transmit a new, fresh pseudonym.

The most commonly found solution in the literature consists of repeatedly applying a hash function to the static identifier (i.e., $pseudonym_n = hash^n(EPC)$). Since the work of Sarma [31] in 2002, there has been a huge number of solutions based on this idea [32–36]. Other authors have proposed using both hash functions and pseudorandom numbers generators [27,29].

The use of pseudonyms may be a good and interesting solution. However, hash functions have not been ratified by the EPC-C1G2 specification. As we have shown above, standard hash functions exceed by far the capabilities of low-cost RFID-tags. Therefore, an interesting issue may be to design efficient and lightweight hash functions for RFID environments.

In another different direction, other authors propose the use of lightweight cryptography. Juels [37] proposal is based on using memory: Tags store a short list of random identifiers or pseudonyms (known by authorized verifiers to be equivalent). Each time the tag is queried, it emits the next pseudonym in the list. On the other hand, Vajda and Buttyán et al. [38–40] proposals are based on using efficient operators (bitwise or, and, xor and sum mod 2^m) to design new ultra-lightweight cryptographic primitives. Although these proposals present some security problems, they stand for a great advance in the research area of securing low-cost RFID tags. In fact, some interesting ideas may be extracted for the incoming Gen-3 specification.

6.6.2 Access Procedures

After acknowledging a tag, a reader may want to access the tag. The access command set comprises *Req_RN, Read, Write, Kill, Lock, Access, Blockwrite,* and *Blockerase.* The above commands can be computed under the following conditions:

Command	State
Req_RN	Acknowledged, open or secured
Read	Secured
Write	Secured
BlockWrite	Secured
BlockErase	Secured
Acess	Open or secured
Kill	Open or secured
Lock	Secured

Note that *Read, Write, Blockwrite,* and *BlockErase* commands also may be executed from the open state when allowed by the lock status of the memory location.

A reader starts an access to a tag in the acknowledged state as set out below:

- $R \Rightarrow T$: The reader sends a *Req_RN* to the acknowledged tag.
- $T \Rightarrow R$: The tag generates and stores a new RN16 (denoted *handle*). Then, the tag backscatters the *handle,* and transitions to open state if its access password is nonzero, or to the secured state if its access password is zero.
- R: Now or sometime later, the reader may start further access commands.

All access commands sent to a tag (in open or secured state) include the *handle* as a parameter. The tag shall verify the *handle* every time an access command is received. Access commands with an incorrect *handle* are ignored. Tag's answers should include the *handle,* whose value is fixed for the entire duration of an access sequence.

Tags and readers can communicate indefinitely in the open and secured states. If the reader wants to finish communication, one of the followings messages would have to be sent: *Query, QueryAdjust, QueryRep,* or *NAK.*

6.6.2.1 Write, Kill, and Access Commands

Everytime a *Write, Kill,* or *Access* command is sent to a tag, a 16-bit word (either data or half-password) is transmitted over the channel. As confidential information is sent, the reader "obscures" it by means of a cover code. In general terms, to cover code data or a password, the reader first requests a random number from the tag.

The reader computes a bitwise *xor* of the word with this random number, and transmits the cover code string to the tag. Finally, the tag uncovers the received word by performing a bitwise *xor* with the original random number. As described in the specification, the following sequence of messages are exchanged:

- $R \Rightarrow T$: The reader sends a *Req_RN* to the acknowledged tag.
- $T \Rightarrow R$: The tag generates a new RN16 and backscatters it to the reader.
- $R \Rightarrow T$: The reader computes a 16-bit cipher-text, which is composed of the bitwise *xor* of the 16-bit word to be transmitted with the new RN16. The reader sends a command to the reader, which includes as a parameter the cipher-text and the *handle*.
- $T \Rightarrow R$: The tag decrypts the received cipher-text performing a bitwise *xor* of it with the original RN16.

Kill and access passwords are 32-bits length. Therefore, to kill or access a tag, a reader shall follow a multistep procedure: The first containing the 16 most significant bits (*MSB*) of the kill or access password *xor*ed with a RN16, and the second containing the 16 least significant bits (*LSB*) of the kill or access password *xor*ed with a different RN16.

The security margin of a protocol using a 16-bit PRNG is usually bounded by $1/2^{16}$. Furthermore, the access and kill password are 32-bit values. The use of 32-bits random numbers would avoid the multistep procedure for using the access, kill, and write command. We recommend the use of 32-bits PRNG, which increases its security level and provides greater flexibility. Therefore, the design of lightweight 32-bits PRNG can be considered as an interesting open issue in RFID security. This PRNG has to obey the randomness criteria specified in the EPC-G1G2 standard [8] and the severe computational and storing requirements for low-cost RFID tags.

The RFID channel asymmetry does not prevent the listening of the backward channel (tag-to-reader channel) by an attacker. Indeed, the security of the above mechanisms can be easily jeopardized even by a passive eavesdropper. Under this scenario, the attacker can obtain the random number (RN16) sent by the tag (answer to *Req_RN* message). Once obtained, the attacker can decrypt the message sent by the reader using a simple *xor*. Therefore, the access password, kill password, or send data can be easily acquired by means of listening to the channel and performing a bitwise *xor*. Summarizing, the cover coding does not provide any kind of security protection to data or passwords.

A naïve solution to protect data and passwords would be the use of conventional encryption. However, standard cryptographic ciphers lie beyond the capabilities of low-cost RFID tags. So, an open issue is the designing of new secure ciphers conforming to the severe restrictions of these environments. Another possible solution may consist on more sophisticated challenge-response protocols, such as the ultra-lightweight mutual authentication protocols proposed by Peris-Lopez et al. in [39,40].

6.6.2.2 Read Command

Read allows a reader to read part or all of the tag's reserved, EPC, TID, or user memory. A read command has the following fields:

- **MemoryBank:** Specifies whether the reader accesses reserved, EPC, TID, or user access memory.
- **WordPtr:** Specifies the starting word address for the memory read, where words are 16-bit in length.
- **WordCount:** Specifies the number of 16-bit words to be read.

The *Read* command also includes the tag's handle and a *CRC*-16. Before receiving a *Read* command, the tag first verifies that all memory words exist and none are read-locked. Second, the tag backscatters a header (a 0-bit), the requested memory words, and its *handle*.

The information exchanged between readers and tags is transmitted in plaintext. So, if the tag requests access to private information, it could be obtained by an attacker listening to the channel. In the case of the access password being equal to zero, the *Reading* command can be accomplished from the open state. This will imply that an adversary may gain access to the tag's data without prior authentication. Once the tag is in open state, an attacker may send messages to the tag trying to obtain all its stored information. To avoid this kind of attack, we recommend locking the memory when the access password is fixed to zero in order to discard *Read* commands.

6.6.2.3 Lock Command

Only tags in the secured state shall execute a *Lock* command. Using the *Lock* command the reader is able to:

- Lock individual passwords, thereby preventing or allowing subsequent reads and/or writes of that password.
- Lock individual memory banks, therefore preventing or allowing subsequent writes to that bank.
- Permalock (make permanently unchangeable): The lock status for a password or memory bank. Permalock bits, once asserted, cannot be deasserted. If a tag receives a *Lock* whose payload attempts to deassert a previously asserted permalock bit, the tag will ignore the command and backscatter an error code.

As displayed in Figure 6.5, *Locks* contains a 20-bit payload defined as follows:

- The first 10 payload bits are *Mask* bits, which are interpreted by tags as follows:

Lock–Command Payload

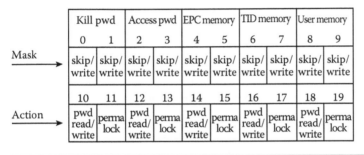

0	1	2	3	4	5	6	7	8	9	10	11	12	13	14	15	16	17	18	19

Kill Mask | Access Mask | EPC Mask | TID Mask | User Mask | Kill Action | Access Action | EPC Action | TID Action | User Action

Mask and Associated Action Fields

	Kill pwd		Access pwd		EPC memory		TID memory		User memory	
	0	1	2	3	4	5	6	7	8	9
Mask →	skip/write	skip/write	skip/write	skip/write	skip/write	skip/write	skip/write	skip/write	skip/write	skip/write
	10	11	12	13	14	15	16	17	18	19
Action →	pwd read/write	perma lock	pwd read/write	perma lock	pwd read/write	perma lock	pwd read/write	perma lock	pwd read/write	perma lock

Figure 6.5   Lock-command payload and masks.

- *Mask* = 0: Ignore the associated action field and retain the current lock setting.
- *Mask* = 1: Implement the associated action field and overwrite the current lock setting.
■ The last 10 payload bits are *Actions* bits, which are interpreted by tags as follows:
  - *Action* = 0: Deassert lock for the associated memory location.
  - *Action* = 1: Assert lock or permalock for the associated memory location.

The functionality of various actions fields is described below:

Pwd-Write	Permalock	Description
0	0	Associated memory bank is writeable from either the open or secured states.
0	1	Associated memory back is permanently writeable from either the open or secured states and may never be locked.
1	0	Associated memory back is writeable from the secured state, but not from the open state.
1	1	Associated memory back is not writeable from any state.

Pwd-read/Write	Permalock	Description
0	0	Associated password location is readable and writeable from either the open or secured states.
0	1	Associated password location is permanently readable and writeable from either the open or secured states and may never be locked.
1	0	Associated password location is readable and writeable from the secured state, but not from the open state.
1	1	Associated password location is not readable or writeable from any state.

The security of the *Lock* command lies in the security of the access password due to the fact that the *Lock* command can only be executed when the tag is in the secured state. For example, a *DoS* attack can be accomplished once the access password is obtained. When the tag is in a secured state, the attacker can send a *Lock* command, whose payload is $F F F F F_h$. This command will entail the permanent lock of the memory, preventing the tag's functionality. Note that the permanent lock is an irreversible action.

6.6.2.4 BlockWrite and BlockErase Commands

As optional commands, readers, and tags may implement *BlockWrite* and *BlockErase*. These commands allow a reader/tag to write/read multiple words in the tag's Reserved, EPC, TID, or User memory using a single command. These commands have the following states:

- **MemoryBank:** Specifies whether the *BlockWrite/BlockErase* accesses Reserved, EPC, TID or User access memory. Both commands apply to a single memory bank. Successive *BlockWrite/BlockErase* can apply to different banks.
- **WordPtr:** Specifies the starting word address for the memory write/read, where words are 16-bit lengths.
- **WordCount:** Specifies the number of 16-bit words to be written/read.
- **Data:** Only employed in the *BlockWrite* command. It contains the 16-bit words to be written, and shall be $16 \times WordCount$ bits in length.

The security of the *BlockWrite* command is also nonexistent. Tags send information (16-bit words) over the channel in plain text. Therefore, if privacy information passes on the channel, it may be captured by a passive attacker listening to the channel. Similarly, an active attacker can easily modify messages.

The *BlockErase* command is not encrypted upon transmission over the channel. As the information is transmitted in clear, an attacker may be able to easily modify the messages. For example, an attacker can send the following message: 0xc8 -01 - *EPC* - 0x00h - 0x8 - *handle*. Once the tag receives the previous message, the EPC memory will be erased, leaving a nonoperative tag.

If the access password is equal to zero, the *BlockWrite* and *BlockErase* command can be accomplished from open state. So, under this situation, an adversary may acquire or erase a tag's data without a previous authentication. It is recommended that these two commands are deactivated when access password is fixed to zero. In fact, a more secure option would be to limit the usage of these two commands, independently of whether a tag is in a secured or open state, only when nonsensitive information is in gear.

6.7 EPC Class-1 Generation-2⁺

The vast majority of work on the designing of security protocols for RFID either do not conform to the EPC-C1G2 specification or suffer from serious security flaws. In this section, some recent works are presented that try to raise the security level of low-cost RFID tags and ones that were developed "conforming" to EPC-C1G2.

6.7.1 Strengthening EPC Tags

Juels [41] claims that EPC tags that do not have any explicit authentication functionality are vulnerable to cloning attacks. Tags emit their EPC in a promiscuous mode and readers accept the validity of the EPC at face value. The result is that tags compliant with EPC-C1G2 are vulnerable to elementary cloning attacks. Juels introduces the concept of skimming, which is the process of scanning a tag by obtaining its EPC for the purpose of cloning.

Several algorithms are proposed, distinguishing between two types of tags. A basic tag is one that carries only the mandatory features of the EPC-C1G2 specification. An enhanced EPC tag additionally includes the optional access-control function.

Following is a description of the *BasicTagAuth* and the *EnhancedTagAuth* protocols. In a system with N tags, the integer i (with $1 \leq i \leq N$) denotes the unique index of an EPC tag. T_i and K_i denote the *EPC* identifier and the valid kill PIN of tag i, respectively. PIN-test(K) denotes an EPC tag command that causes a tag to output a bit-response b: 0 if the kill PIN is correct, but outputs 1 otherwise.

6.7.1.1 Basic TagAuth Protocol

```
T:        T ← T_i
T ⇒ R:    T
R:          if T = T_x for some 1 ≤ x ≤ N the i ← x
              else output "unknown tag" and halt
R:          (j, {P_i^(1), P_i^(2),..., P_i^(q)}) ← GeneratePINSet(i) [q]
            M ← valid
          for n = 1 to q do
R ⇒ T:      PIN-test(P_i^(n))
T ⇒ R:      b
            if b = = 1 and n ≠ j then M ← invalid
            if b = = 0 and n = j then M ← invalid
R:         output M;
```

The key idea of this protocol is the presentation of spurious PINs as a means of testing their authenticity. The q value is a security parameter that specifies the number of spurious PINs to be generated.

An attacker that performs skimming attacks can only create a cloned device that attempts to guess the correct PIN-trial j uniformly at random. The probability of a successful attack (the cloned tag appearing to be valid) is clearly just $1/q$. However, the protocol is very inefficient (time-consuming) as large number of messages (q) are exchanged between tags and readers to provide an accurate security level.

6.7.1.2 Enhanced TagAuth Protocol

```
T:        T ← T_i
T ⇒ R:    T
R:        if T = T_x for some 1 ≤ x ≤ N the i ← x, A ← A_i
                else output "unknown tag" and halt
R ⇒ T:    A
T:        if A = A_i then K ← K_i
                else K ← ϕ
T ⇒ R:    K
T:        if K = K_i then output "valid"
                else output "invalid"
```

Enhanced EPC tags have both access and kill PINs. Juels proposes a mutual authentication protocol in which the access PIN (A_i) serves to authenticate the reader and the kill PIN (K_i) in turn serves to authenticate the tag.

Both of Juels' proposals prevent a cloned tag from impersonating legitimate EPC-C1G2 tags. Although this problem constitutes a very interesting issue, the presented proposals are only resistant to skimming attacks, being vulnerable to nearly all others: eavesdropping, active attacks, physical attacks, etc. Furthermore, we believe that focusing only on cloning and forgetting other important problems, such as privacy, tracking, denial of service, etc., should not be the correct approach.

6.7.2   Shoehorning Security into the EPC Standard

Juels and Bailey [42] examine various ways that RFID tags might perform cryptographic functionality while remaining compliant with the EPC-C1G2 standard. Their key idea resides in taking an expansive view of EPC tag memory. Instead of considering this memory merely as a form of storage, they use it as an input/output medium for interfacing with a cryptographic module within the tag. Therefore, read/write commands may carry out cryptographic values associated, such as messages in a challenge-response protocol.

Juels and Bailey claim that EPC-C1G2 is a very limited protocol for entity authentication. However, tags are not authenticated in the specification, which facilites counterfeiting. As an example, this simple challenge-response protocol is suggested:

- $R \Rightarrow T: C_R$
- $T \Rightarrow R: EPC, R_T$

$R_T = H(K_{TS}, C_R)$, and $H()$ is a cryptographic primitive like a block-cipher, K_{TS} is some secret key known only to the tag and the reader, and C_R is a unique challenge.

To implement the previous protocol, commands designed for other purposes will have to be reused or one should define custom or new commands. The task of defining the use of one protocol to carry the data units of another is often designed as protocol convergence. Concisely, the ISO 7816 command set to accomplish entity authentication of the tag is proposed.

Juels and Bailey's work signals the need for mutual authentication between tags and readers. However, the proposal is based on the assumption that EPC-C1G2 tags might support onboard cryptographic modules. This is probably not realistic, at least at the present time. Moreover, this proposal is focused on tag authentication (counterfeiting). However, there are a great number of other security concerns that should be considered before proposing a protocol convergence. Otherwise, Juels' and Bailey's proposal can be a good starting point.

6.7.3   Enhancing Security of EPC-C1G2

Duc et al. [43] propose a tag-to-backend server authentication protocol. During manufacturing, EPC and a tag's access PIN are assigned. Then, it chooses a random seed and store $K_1 = f(seed)$ in the tag's memory and the corresponding backend server's database entry. The authentication protocols is described as follows, where *f*() denotes a PRNG function.

$R \Rightarrow T$:   Query request

T: Compute $M_1 = CRC(EPC||r) \oplus K_i$ and $C = CRC(M_1 \oplus r)$, where r is a nonce.

$T \Rightarrow R \Leftrightarrow S$: M_1, C and r

S: From each tuple (EPC, K_i) in its database, the server verifies whether the equation $M_1 \oplus K_i = CRC(EPC||r)$ and $C = CRC(M_1 \oplus r)$ hold. If a match is found, then the tag is successfully identified and authenticated, and the server will forward a tag's information to the reader and proceed to the next step; otherwise the process is stopped.

$S \Leftrightarrow R \to T$ M_2

If R desires to perform read/write operation to a tag's memory, it requests an authentication token to M_2 from S where $M_2 = CRC(EPC||PIN||r) \oplus K_i$. Then, R sends M_2 to T. The tag receives M_2 and computes its M_2, using its local values (PIN, r, EPC, K_i), and verifies whether the received M_2 equals the local one. If so, the tag will accept the end session command in the next step.

$S \leftarrow R \to T$:   *EndSession* command

Upon receiving this command, both server and tag update their shared keys as $K_{i+1} = f(K_i)$.

The security of Duc et al.'s protocol greatly depends on key synchronization between tags and backend server. The last message of the protocol is comprised of an *EndSession* command, which is sent to both tags and readers. An interception of one of these messages will cause a synchronization loss between the tag and the server. Therefore, the tag and the server cannot authenticate each other any more. Additionally, this fault may be exploited as follows: The *EndSession* message to server may be intercepted avoiding its key updating and then a counterfeiting tag can replay old data (M_1, r, C) leading to its correct authentication.

Furthermore, the forward secrecy is not guaranteed. Forward secrecy is the property that guarantees that the security of the messages sent today will be valid tomorrow. If the tag is compromised, obtaining (EPC, PIN, K_i), an attacker can verify whether past communications came from the same tag. This attack is based in the symmetry of messages M_1 and M_2. Imagine that an attacker has stored old (M_1, M_2, r) messages. Now, he computes $M_1 \oplus M_2 = CRC(EPC \oplus r) \oplus CRC(EPC||PIN||r)$, and using the compromised values (EPC, PIN, K_i) and the eavesdropped r, he could verify that these messages came from the same tag.

6.7.4 Mutual Authentication Protocol

Chien and Chen [44] proposed a mutual authentication protocol in order to solve the security weakness of Duc et al.'s proposal. The scheme consists of two phases: initialization and authentication.

6.7.4.1 Initialization Phase

For each tag denoted as T_i, the server randomly selects an initial authentication key K_{i_0} and an initial access key P_{x_0}. These two values joined with the EPC (EPC_i), are stored in the tag. The authentication key and the access key will be updated after each successful authentication. For each tag, the server maintains a record of six values in its database: (1) EPC_i; (2) the old authentication key for this tag (K_{old}), and is initially set to K_{i_0}; (3) P_{old} denotes the old access key for this tag, and is initially set to P_{i_0}; (4) K_{new} denotes the new authentication key, and is initially set to K_{i_0}; (5) P_{new} denotes the new authentication key, and is initially set to P_{i_0}; and (6) Data denotes all the information about the tagged object.

6.7.4.2 The (n+1) Authentication Phase

$R \Rightarrow T_i$: N_1

The reader sends a random nonce N_1 as a challenge to the tag.

$T_i \Rightarrow R \Rightarrow S$: M_1, N_1, N_2

The tag generates a random number N_2, computes $M_1 = CRC(EPC_i || N_1 || N_2) \oplus K_{i_n}$, and sends the value back to the reader, which will forward these values to the server.

The server interactively picks up an entry (EPC_i, K_{old}, K_{new}, P_{old}, P_{new}, Data) from its database, computes $I_{old} = M_1 \oplus K_{old}$ and $I_{new} = M_1 \oplus K_{new}$; and checks whether any of two equations $I_{old} = CRC(EPC_i || N_1 || N_2)$ and $I_{new} = CRC(EPC_i || N_1 || N_2)$ hold. The process is repeated until a match is found, meaning a successfully authentication of the tag. If a match is not found, a failure message is sent to the reader and the process is stopped.

$S \Rightarrow R$: M_2, Data

After a successful authentication, the server computes $M_2 = CRC(EPC_i || N_2) \oplus P_{old}$, or $M_2 = CRC(EPC_i || N_2) \oplus P_{new}$, depending on which value (K_{old}, K_{new}) satisfies the verification equation in the previous step. It also updates $K_{old}, = K_{new}$, $P_{old}, = P_{new}$, $K_{new} = PRNG(K_{new})$, and $P_{new} = PRNG(P_{new})$. The server sends M_2, Data to the reader.

$$R \Rightarrow T_i: \qquad M_2$$

> Upon receiving M_2, the tag verifies whether the equation $M_2 \oplus P_{i_n} = CRC(EPC_i || N_2)$ holds. If so, it updates its keys as $K_{i_{n+1}} = PRNG(K_{i_n})$ and $P_{i_{n+1}} = PRNG(P_{i_n})$.

Chien and Chen claim that the two sets of authentication keys and access keys provide a defense against DoS attacks that may cause desynchronization between the tag and the reader. However, a malicious reader can easily accomplish an attack causing loss of synchronization. The attacker sends a nonce N_1 to the objective tag. The tag backscatters M_1 and N_2. Now, the attacker has to generate M_2. According to EPC-C1G2 specification, tags support onboard a 16-bit CRC, so the message M_2 is a 16-bit length. Although the attacker does not know any tag's secret information, an exhaustive search can be carried out. A tag compliant with EPC-C1G2 standard has to provide nearly 500 answers/sec (real conditions where noise may exist). Under these conditions, an exhaustive search of M_2 takes approximately two minutes. So the attacker only has to send all these messages to the tag, and one of those will lead to the update of the tag's key. At this moment, the tag and the database can no longer be authenticated. Therefore, Chien and Chen's protocol is not immune to denial-of-service attacks, either.

Moreover, Peris-Lopez et al. [45] have recently presented a deeper security analysis of Chien and Chen's protocol. From this analysis, it can be concluded that none of the authentication protocol objectives are met. Unequivocal identification of tagged items is not guaranteed due to birthday attacks. Furthermore, an attacker can impersonate not only legitimate tags, but also the backend database. Location privacy is easily jeopardized by a straightforward tracking attack. Additionally, a successful auto-desynchronization (DOS attack) can be accomplished in the backend database despite the security measures taken against it. Many of these results are a direct consequence of the linearity of cyclic redundancy codes (CRCs). The disaster of wired equivalent privacy (WEP) [46] and Secure SHell (SSH) [47,48] protocols are well-known examples of similar mistakes.

6.8 Conclusions

This chapter has presented, in some detail, the EPCglobal Class-1 Generation-2 specification. This specification may be considered as a "universal standard" for low-cost RFID tags after its ratification by both EPCglobal and ISO.

First, the main concepts of the standard have been explained in order to introduce the reader to the subject. Second, a security analysis of all its commands (mandatory and optional) has been accomplished. From this examination, we can conclude that the security level of EPC-C1G2 specification is very weak. Protection against security problems, such as privacy and tracking, is not guaranteed. Although tags possess an onboard PRNG, it is mainly used to establish an identifier session, not for security reasons. In order to increase the security level and avoid the multistep procedures, we recommend that tags support an onboard 32-bit PRNG.

Finally, some recent works that try to raise the security level of low-cost RFID tags "conforming" to the EPC-C1G2 have been presented. However, these protocols have important security pitfalls, as it has been shown. Some of the proposals ask the CRC checksums for properties only found on standard hash-functions, something which is not realistic and generates the possibility of many attacks. It also leaves open the issue of designing efficient, lightweight hash functions. Another related open issue is the designing of ciphers suitable to these constrained environments. These protocols are based on several keys shared between the tag and readers (database). Improvements to key management are another thought-provoking issue that should be considered by researchers.

To conclude, we expect that in the future EPC Generation-3 standard, a higher security level will be guaranteed. In this work, we have provided many ideas and insights to help in solving the security pitfalls of its predecessor.

References

[1] "Frequently asked questions". http://www.rfidjournal.com, 2006.

[2] C.M. Roberts. Radio frequency identification (RFID). *Computers and Security*, 25(1):18–26, 2006.

[3] M. Rieback, B. Crispo, and A. Tanenbaum. The evolution of rfid security. *IEEE Pervasive Computing*, 5(1):62–69, 2006.

[4] RFID Consultation Website. http://www.rfidconsultation.eu/. May 15, 2006.

[5] P. Peris-Lopez, J. Cesar Hernandez-Castro, J. Estevez-Tapiador, and A. Ribagorda. RFID systems: A survey on security threats and proposed solutions. *In Proc. of PWC06*, vol. 4217 of Lecture Notes in Computer Science, Springer-Verlag, Heildeberg, Germany, pp. 159–170, 2006.

[6] EPCglobal. http://www.epcglobalinc.org/. February 5, 2006.

[7] ISO — International Organization for Standardization. http://www.iso.org/. February 15, 2006.

[8] Class-1 Generation-2 UHF air interface protocol standard, version 1.0.9: "Gen-2." http://www.epcglobalinc.org/, January 2005.

[9] EPCglobal Class-1 Gen-2 RFID Specification. Alien Technology, Morgan Hill, CA, whitepaper, 2006.

[10] M. Weiser. The computer for the 21st century. *Scientific American*, 265(3):94–104, September 1991.

[11] S. Weis, S. Sarma, R. Rivest, and D. Engels. Security and privacy aspects of low-cost radio frequency identification systems. In *Proc. of SPC'03*, vol. 2802 of Lecture Notes in Computer Science, Springer-Verlag, Heildeberg, Germany, pp. 454–469, 2003.

[12] D. Ranasinghe, D. Engels, and P. Cole. Low-cost RFID systems: Confronting security and privacy. In *Auto-ID Labs Research Workshop*, Zurich, Switzerland, September, 2004.

[13] M. Feldhofer and C. Rechberger. A case against currently used hash functions in RFID protocols. In *Proc. of RFIDSec'06*, Graz, Austria, July 2006.

[14] K. Yksel, J.P. Kaps, and B. Sunar. Universal hash functions for emerging ultra-low-power networks. In *Proc. of CNDS'04*, San Diego, CA, January 2004.

[15] Amphion: CS5265/75 AES Simplex encryption/decryption. http://www.amphion. com, March 10, 2005.

[16] M. Feldhofer, K. Lemke, E. Oswald, F. Standaert, T. Wollinger, and J. Wolker-storfer. State of the art in hardware architectures. In *Technical Report, ECRYPT Network of Excellence in Cryptology*, 2005.

[17] M. Feldhofer, J. Wolkerstorfer, and V. Rijmen. AES implementation on a grain of sand. In *Proc. of IEEE on Information Security*, IEEE Computer Society, vol. 152, pp. 13–20, 2005.

[18] T. Good, W. Chelton, and M. Benaissa. "Review of stream cipher candidates from a low resource hardware perspective." http://www.ecrypt.eu.org/stream/hw.html, March 2006.

[19] M. Hell, T. Johansson, and W. Meier. "Grain: A stream cipher for constrained environments and Grain-128." http://www.ecrypt.eu.org/stream/, September 10, 2006.

[20] C. de Canniere and B. Preneel. "Trivium specifications." http://www.ecrypt. eu.org/ stream/, October 10, 2006.

[21] O. Kucuk. "Slide resynchronization attack on the initialization of Grain 1.0." http://www.ecrypt.eu.org/ stream/, October 15, 2006.

[22] H. Gilbert, C. Berbain, and A. Maximov. "Cryptanalysis of grain." http://www. ecrypt.eu.org/ stream/, September 20, 2006.

[23] M. Hell, T. Johansson, and W. Meier. Grain—a stream cipher for constrained enviroments. In *Proc. of RFIDSec'05*, 2005, Graz, Austria, July, 2005.

[24] S.A. Weis, S.E. Sarma, R.L. Rivest, and D.W. Engels. Security and privacy aspects of low-cost radio frequency identification systems. In *Proc. of Security in Perva-sive Comp.*, vol. 2802 of Lecture Notes in Computer Science, Springer-Verlag, Heildeberg, Germany, pp. 201–212, 2004.

[25] D. Molnar and D. Wagner. Privacy and security in library RFID: issues, practices, and architectures. In *Proc. of ACM CCS'04*, pp. 210–219. ACM, ACM Press, New York, October 2004.

[26] C. Chatmon, T. Van Le, and M. Burmester. Secure anonymous RFID authentica-tion protocols. In *Technical Report TR-060112*, 2006.

[27] T. Dimitriou. A lightweight RFID protocol to protect against traceability and cloning attacks. In *Proc. of PerCom'06*, Pisa, Italy, 2006.

[28] S. Lee, T. Asano, and K. Kim. RFID mutual authentication scheme based on syn-chronized secret information. In *Proc of Symposium on Applied Cryptography and Information Security*, Glasgow, Scotland 2006.

[29] K. Rhee, J. Kwak, S. Kim, and D. Won. Challenge-response based RFID authenti-cation protocol for distributed database environment. In *Proc. of SPC'05*, vol. 3450 of Lecture Notes in Computer Science, Springer-Verlag, Heildeberg, Germany, pp. 70–84, 2005.

[30] G. Tsudik. YA-TRAP: Yet another trivial RFID authentication protocol. In *Proc. of PerCom'06*, Pisa, Italy, 2006.

[31] S.E. Sarma, S.A. Weis, and D.W. Engels. RFID systems and security and privacy implications. In *Proc. of CHES'02*, vol. 2523 of Lecture Notes in Computer Science, Springer-Verlag, Heildeberg, Germany, pp. 454–470, 2002.

[32] E.Y. Choi, S.M. Lee, and D.H. Lee. Efficient RFID authentication protocol for ubiquitous computing environment. In *Proc. of SECUBIQ'05*, vol. 3823 of Lecture Notes in Computer Science, Springer-Verlag, Heildeberg, Germany, pp. 945–954, 2005.

[33] D. Henrici and P. Müller. Hash-based enhancement of location privacy for radio-frequency identification devices using varying identifiers. In *Proc. of PERSEC'04*, pp. 149–153. IEEE Computer Society, Washington, D.C., 2004.

[34] S.M. Lee, Y.J. Hwang, D.H. Lee, and J.I.L. Lim. Efficient authentication for low-cost RFID systems. In *Proc. of ICCSA'05*, vol. 3480 of Lecture Notes in Computer Science, Springer-Verlag, Heildeberg, Germany, pp. 619–627, 2005.

[35] M. Ohkubo, K. Suzuki, and S. Kinoshita. Cryptographic approach to "privacy-friendly" tags. In *Proc. of RFID Privacy Workshop*, 2003.

[36] J. Yang, J. Park, H. Lee, K. Ren, and K. Kim. Mutual authentication protocol for low-cost RFID. In *Proc RFIDSec'05*, 2005.

[37] A. Juels. Minimalist cryptography for low-cost RFID tags. In *Proc. of SCN'04*, vol. 3352 of Lecture Notes in Computer Science, Springer-Verlag, Heildeberg, Germany, pp. 149–164, 2004.

[38] I. Vajda and L. Buttyán. Lightweight authentication protocols for low-cost RFID tags. In *Proc. of UBICOMP'03*, Atlanta, GA. 2003.

[39] P. Peris-Lopez, J.C. Hernandez-Castro, J. Estevez-Tapiador, and A. Ribagorda. M2AP: A minimalist mutual-authentication protocol for low-cost RFID tags. In *Proc. of UIC'06*, vol. 4159 of Lecture Notes in Computer Science, Springer-Verlag, Heildeberg, Germany, pp. 912–923, 2006.

[40] P. Peris-Lopez, J.C. Hernandez-Castro, Juan M. Estevez-Tapiador, and A. Ribagorda. EMAP: An efficient mutual authentication protocol for low-cost RFID tags. In *Proc. of IS'06*, vol. 4277 of Lecture Notes in Computer Science, Springer-Verlag, Hcildeberg, Germany, pp. 352–361, 2006.

[41] A. Juels. Strengthening EPC tags against cloning. March 2005.

[42] A. Juels and D. Bailey. Shoehorning security into the EPC standard In *Proc. of SCN'06*, vol. 4116 of Lecture Notes in Computer Science, Springer-Verlag, Heildeberg, Germany, pp. 303–320, 2006.

[43] D.N. Duc, J. Park, H. Lee, and K. Kim. Enhancing security of EPCglobal Gen-2 RFID tag against traceability and cloning. In *Proc. of Symposium on Cryptography and Information Security*, January, 2006.

[44] H.Y Chien and C.H Chen. Mutual authentication protocol for RFID conforming to EPC Class-1 Generation-2 standards. *Computer Standards and Interfaces*, 29(2):254–259, 2007.

[45] P. Peris-Lopez, J.C. Hernandez-Castro, J. Estevez-Tapiador, and A. Ribagorda. Cryptanalysis of a novel authentication protocol conforming to EPC-C1G2 standards. In *Proc. of RFIDSec'07*, Malaga; Spain, July 2007.

[46] N. Borisov, I. Goldberg, and D. Wagner. Intercepting mobile communications: the insecurity of 802.11. In *Proc. of MobiCom'01, ACM*, ACM Press, New York, pp. 180–189, 2001.

[47] A. Futoransky, and E. Kargieman. An attack on CRC-32 integrity checks of encrypted channels using CBC and CFB modes. http://www.coresecurity.com/files/attachments/CRC32.pdf, 1999.

[48] N. Provos, and P. Honeyman. ScanSSH—Scanning the Internet for SSH servers. In *Proc. of USENIX'01*, San Diego, CA, December 2001.

Chapter 7

RFIG: Geometric Context of Wireless Tags

Ramesh Raskar, Paul Beardsley,
Paul Dietz, and Jeroen van Baar

Contents

Radio frequency tags allow objects to become self-describing, communicating their identity to a close–at-hand radio frequency (RF) reader. Our goal is to build a Radio Frequency Identification and Geometry (RFIG) transponder. In addition to identity, a RFIG tag can record and respond its own geometric context, such as absolute and relative location with respect to adjacent RFIG tags.

7.1   Geometric Context of Wireless Tags

The geometry-rich functionality is achieved by augmenting each tag (Figure 7.1) with a photosensor (Figure 7.2). Optical communication with this composite RF-photosensing tag is achieved using modulated light. In this chapter, the

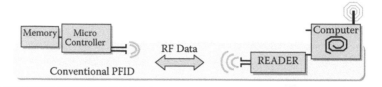

Figure 7.1 Conventional Radio Frequency Identifaction (RFID) transponder communicates with RF reader and responds with the id number stored in the tag's memory.

operations are shown using a projector that is paired with the tag RF reader. The projector performs the dual operation of sending optical data to the tag (similar to a television's infrared (IR) remote control unit) and also giving visual feedback by projecting instructions onto objects. Current tag readers operate in broadcast mode with no concept of a directional communication; however, the RFIG tags allow the locating of tags within a few millimeters, support selection of individual tags, and create a two dimensional (2D) or three dimensional (3D) coordinate frame for the tags. The system of projector and photosensing tags offer a set of rich, geometric operations. It presents a new medium for many of the results from the area of computer vision, with projector and tags replacing camera and image interest points (Raskar, Beardsley et al., 2004; Raskar, Nii et al., 2007).

The experimental work in this chapter is based on active, battery-powered RF tags. However, our goal has been to develop methods that can be used with passive, nonpowered Radio Frequency IDentification (RFID) tags. The key issue in evolving this active tag system to passive tags would be "power." In our work, we only allowed computation and sensing consistent with the size and power levels we felt were achievable on a passive RFID system. For example: (1) tags are not photosensing or computing until woken by the RF reader and (2) there isn't a light-emitting diode (LED) on the tag as a visual beacon to a human or camera-based system because it would be power-hungry.

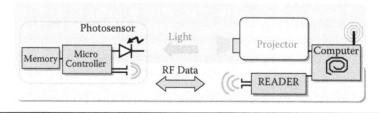

Figure 7.2 The RFIG tag: Radio frequency identity and geometry (RFIG) transponder communicates with RF reader as well as spatiotemporal light modulator, such as a modulated infrared (IR) light. For example, with a full-fledged data projector, one can find the stored id as well as the (x,y) projector pixel location that illuminates the tag.

Location tracking using RF-received signal strength or time of arrival is popular, but requires multiple readers, and the accuracy may be insufficient for complex geometric procedures (Hightower and Borriello, 2001). Previous systems also have married RF tags with optical or ultrasound sensors to improve accuracy. Some systems use active RF tags that respond to laser pointers. The FindIT Flashlight project uses a one-way interaction and an indicator light on the tag to signal that the desired object has been found (Ma and Paradiso, 2002). Other systems use a two-way interaction where the tag responds back to the personal digital assistant (PDA) using a power-hungry protocol, such as the 802.11 or X10 (Patel and Abowd, 2003; Ringwald 2002). Cooltown (Hewlett Packard*) (The Cooltown Project, 2001) uses beacons that actively transmit device references, but without the ability to point and without visual feedback. The Cricket project (Teller et al., 2003) recovers PoseAware of a handheld device using installed RF and ultrasound beacons as well as doing recovers position and orientation of a handheld device.

7.1.1 How It Works

Conventional tag communication works by broadcast from a RF reader, with response from all in-range tags. Limiting the communication to a required tag is traditionally achieved using a short-range tag reader and close physical placement with respect to the tag. In contrast, we can select tags for interaction from a longer-range using projected light, while ignoring unwanted in-range tags. The handheld device first transmits a RF broadcast. Each in-range tag is woken by the signal, and its photosensor takes a reading of ambient light, which is to be used as a zero for subsequent illumination measurements. The projector illumination is turned on. Each tag that detects an increase in illumination sends a response to indicate that it is in the beam of the projector and is ready for interaction.

The handheld device is aimed casually in the direction of a tagged surface, and then sends a RF signal to synchronize the tags, followed by illumination with a sequence of binary patterns, i.e., binary structured light. Each projector pixel emits a unique temporal Gray code and, thus, encodes its position. The tag records the Gray code that is incident on its photosensor, and then makes a RF transmission of its identity, plus the recorded Gray code back to the RF reader. The projector uses the identity plus the recorded (x,y) location to project instructions, text, or images on the tagged object. It is then straightforward to create correctly positioned augmentation on the tagged surface.

7.1.2 Applications

Several aspects of RFIG have been described in our previous work (Raskar et al., 2004). The work was motivated in terms of the commercially important application of inventory control. But, we believe that photosensing tags may have many innovative uses, and, in this chapter, our goal is to present the new ideas in the context of a few promising examples. Below is outlined a broad mode of deployment for geometric analysis. Note that these are speculative uses, not actual work done.

1. Location Feedback, e.g., warehouse management (Figure 7.3): Consider the task of locating boxes containing perishable items about to expire. Even with traditional RF tagging with expiry date information recorded in the indexed database, the employee would have to serially inspect boxes and mark the about-to-expire product boxes. Using RFIG tags, the handheld or fixed projector first locates the queried tags and then illuminates them with symbols, such as *X* and *OK,* so that the employee has a visual feedback. Note that a second user can perform similar operations without RF collision with the first reader or the tags because the two projector beams do not overlap.

2. Obstruction Detection, e.g., object obstructing a railroad track (Figure 7.4): A common computer vision task with a camera includes detecting abnormal conditions by performing image processing. One example is detecting obstruction on railway tracks; for example, raising an alarm if a person or some suspicious material is on subway tracks. Processing images of videos from camera-based systems to detect such events is difficult because the ambient lighting conditions can change and several other activities can result in false positives. But, one can

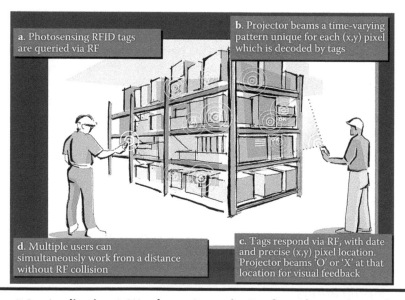

Figure 7.3 Application 1, Warehouse Scenario: Employee locates items about to expire and gets visual feedback. A second user performs similar operations, without conflict in the interaction because the projector beams do not overlap.

Figure 7.4 **Application 2, Detecting obstruction on railway tracks, such as person on tracks near a platform, disabled vehicle at a railroad intersection, or suspicious material on tracks. Finding obstruction with a camera-based system is difficult. The idea is to sprinkle RFIG tags along the track and illuminate them with a fixed or steered beam of temporally modulated light (not necessarily a projector). Tags respond with status of reception of the modulated light. Lack of reception indicates an obstruction, which can be relayed to a central monitoring facility where a human observer can carefully observe the scene, possibly with a pan/tilt/zoom surveillance camera.**

solve this vision problem by sprinkling RFIG tags along the track. These tags can be illuminated with a fixed or steered beam of temporally modulated light (not necessarily a projector), such as a 40 kHz infrared beam from a sparse array of light emitters. Then, the operation is similar to the "beam break" technique commonly used to detect intruders. But, a wireless tag-based system is ideally suited for applications, where running wires to both ends is impractical. Using retro-reflective markers and detecting a return beam is another common strategy to avoid wires, but sprinkling a large number of markers creates an authoring nightmare. In the case of RFIG, the tags identification and location can be easily reported along with the status of reception of the modulated light. Lack of reception indicates obstruction, which can be relayed to a central monitoring facility where a human observer can carefully observe the scene possibly with a pan/tilt/zoom surveillance camera.

3. Ordered Placement and Orientation, e.g., books in a library (Figure 7.5): A common task in libraries, pharmacies, or for facility managers is maintaining a large number of objects in a predetermined order. For example, in a library, if books are RF tagged, it is easy to obtain a book list within the RF range. However, without location information, it is difficult to find out which books

Figure 7.5 Application 3, Books in a library. If books are RF tagged, it is easy to obtain a list of books within the RF range. However, without location information, it is difficult to find out which books are out of alphabetical order. In addition, without book orientation information, it is difficult to detect books whose spines are placed upside down. With RFIG and a handheld projector, the system finds a book title as well as location. Based on the mismatch in title sort with respect to the location sort, the system gives immediate visual feedback and instructions, shown as arrows for original positions.

are out of alphabetical order. In addition, without book orientation information, it is difficult to detect books that are placed with spines upside down. With RFIG and a handheld projector, the system lists the book title as well as location. Then, the system sorts books by title as well as by their 2D geometric location. A mismatch in the two sorted lists indicates that the corresponding book is placed in a wrong position. The system knows the current location for the books as well their ideal position. The projector display gives immediate visual feedback and instructions (shown in Figure 7.5 as arrows from current positions to intended positions). A single book also can be tagged with two RFIG transponders, one at the top of the book spine and one at the bottom. Then, comparing the coordinates of these two tags allows one to find out if the book has been placed upside down.

4. 3D Path Planning/Guiding, e.g., guiding a robot on an assembly line for arbitrarily oriented objects (Figure 7.6): RFIG tags can be used in factories for robot guidance. The idea is similar to other "laser-guided" operations. Suppose a robot is instructed to grab a certain object from a pile moving on a conveyor belt. RFID can simplify the object recognition problem in the machine's vision, but precisely locating the object would be difficult. The idea is to use a fixed projector to first locate the RFIG-tagged object and then illuminate the object with a steady, easily identifiable temporal pattern. A camera attached to the robot arm locks onto this pattern by doing pattern matching and allows the robot to home in on the object.

Figure 7.6 Application 4, Laser-guided robots: To guide a robot to pick a certain object from a pile moving on a conveyor belt, the projector locates the RFIG-tagged object and illuminates it with an easily identifiable temporal pattern. A camera attached to the robot arm locks onto this pattern and allows the robot to home in on the object.

Notice that in a majority of the applications described above, the projector behaves similar to devices that we are all familiar with, e.g., remote controls and laser pointers, but with some spatial or temporal modulation of light. The projector is a glorified remote control communicating with a photosensor in the location-sensing phase and a glorified laser pointer in the image projection phase.

Discussion

Several problems can influence optical communication between the projector and a tag. It can be affected by ambient light; however, wavelength division multiplexed communication is commonly used to solve this problem (e.g., TV remote and IR photosensor). The optical communication also gets noisier as projector–tag distance increases, and if the photosensor gets dirty. However, within these limitations, the RFIG method can support very intricate and multipurpose geometric operations with the ambient intelligence provided by wireless tags. The work indicates some of the possibilities for blurring the boundaries between the physical and digital worlds by making the everyday environment into a self-describing wireless data source, a display surface, and a medium for interaction.

References

Cooltown Project, The. 2001. http://www.cooltown.com/research/.

Hightower, J. and Borriello, G. Location systems for ubiquitous computing. *Computer*, 34(8): 57–66, August 2001.

Ma, H. and PARADISO, J. A. 2002. The FindIT flashlight: Responsive tagging based on optically triggered microprocessor wakeup. *Ubicomp 2002 International Conference*, Goteborg, Sweden, Springer-Verlag, Heidelberg, Germany, pp 160–167.

Patel, S.N. and Abowd, G. D. 2003. A 2-way laser-assisted selection scheme for handhelds in a physical environment. *Ubicomp International Conference 2003*, Seattle, WA, October 2003, Springer-Verlag, Heidelberg, Germany, pp. 200–207.

Raskar, R., Beardsley, P.,Van Baar, J., Wang, Y., Dietz, P., Lee, J., Leigh, D. and Willwacher, T. 2004. RFIG lamps: Interacting with a self-describing world via photosensing wireless tags and projectors. *ACM Trans. Graph. (SIGGRAPH) 22*, 3: 809–818.

Raskar, R., Nii, H., deDecker, B., Hashimoto, Y., Summet, J., Moore, D., Zhao, Y., Westhues, J., Dietz, P., Barnwell, J., Nayar, S., Inami, M., Bekaert, P., Noland, M., Branzoi, V., and Bruns, E. 2007. Prakash: lighting aware motion capture using photosensing markers and multiplexed illuminators. ACM Trans. Graph. 26, 3 (Jul. 2007), 36. DOI=http://doi.acm.org/10.1145/1276377.1276422.

Ringwald, M. 2002. Spontaneousinteraction with everyday devices using a PDA workshop on supporting spontaneous interaction in ubiquitous computing settings. *UbiComp 2002 International Conference*, Goteborg, Sweden, Springer-Verlag, Heidelberg, Germany.

Teller, S., Chen, J., and Balakrishnan, H. 2003. Pervasive poseaware applications and infrastructure. *IEEE Computer Graphics and Applications*, July/August 2003, Canmore, Alberta, Canada.

Chapter 8

RFID Application in Animal Monitoring

Vasileios Ntafis, Charalampos Z. Patrikakis,
Eirini G. Fragkiadaki, and Eftychia
M. Xylouri-Fragkiadaki

Contents

8.1 Introduction

Use of Radio Frequency IDentification (RFID) technology in animal tracking has been a practise for the past number of years in countries around the world. RFID technology is used to track domestic, wild, and farm animals, during which there have been efforts to standardize the methods and the specific RFID tag technology used. As examples of the above, a U.S. Department of Agriculture (USDA) project is in effect for tracking captive deer and elk in the United States based on the use of RFID tags in order to determine how deer and elk contract chronic wasting disease, a fatal neurological disorder. In the standardization field, the U.S. Animal Health Association has already adopted a draft U.S. Animal Identification Plan developed by a group of representatives from the cattle industry and state and federal agencies overseeing the food supply. At the same time, AIM Global has developed a draft standard for RFID in food animals to address growing concerns about the threat of terrorist attacks and the recent outbreaks of both bovine spongiform encephalopathy (BSE: mad cow disease) and hoof-and-mouth disease in different parts of the world.

As the use of RFID technology gets less expensive, more countries are adopting the use of animal tagging. As an example, the USDA has announced plans for the universal tagging of livestock in the United States with RFID by the year 2009.

Providing a brief but comprehensive coverage of the use of RFID technologies in animal identification, tagging, and monitoring, this chapter presents the use of RFID technology for supporting all animal-related RFID activities. A reference to the current standardization and coverage of the use of RFID technology, as defined through global organizations and associations, will be provided as an introductory section to these issues. Following this approach, the chapter will provide a comprehensive coverage of the use of RFID in relation to domestic, farm, and wild animals, the particularities that this use has in each case, while, on the other hand, providing example prototype uses for each case. Legislation issues, that are addressed globally, will also be tackled, giving an overview of the existing legislative framework for the use of RFID as regards to animals. Note that emphasis is placed only on RFID technology as it is used for monitoring animal activities. Therefore, basic knowledge of the use and operation of RFID systems is required, even though the chapter does not address technical issues of RFID.

8.1.1 Systems of Identification and Traceability

The necessity of systems of identification and traceability becomes more obvious day by day. The demands of quality control in agricultural animals, as well as technology's evolution, comprise these two factors that promote the ceaseless improvement of the identification systems. Only few countries worldwide have managed to establish such systems. National laws and regulations come constantly into force, laying new guidelines for these systems so as to improve their credibility. At the same time, technology has reached a point where the use of electronic means of identification (RFID) is possible, resulting in the increasing use of the application.

The outbreak of BSE, commonly known as mad cow disease, revealed the necessity of systems of identification and traceability. During the last two decades, concerns about the safety and quality of food have increased at both governmental and consumer levels. The importance of traceability of animals and animal products has grown as food production and marketing have been removed from direct consumer control. Worldwide trading increases the possibilities of a disease outbreak, when at the same time it makes food traceability more difficult. Despite these, only little progress has been achieved in the field of the systems mentioned above (McKean, 2001; Saa et al., 2005).

In the European community, there are regulations for agricultural animal's traceability and movements. Thus, detection of animals during outbreaks is possible. Moreover, in many cases, countries give subsidies depending on the number of animals in a herd and other requirements, to farms with cattle, sheep, and goats. As the determination of the number is difficult, there is always the possibility of fraud. Good identification systems can help minimize this fraud (Wismans, 1999).

Therefore, with the application of identification systems, animal diseases (sometimes foreign for a country) can be controlled, surveyed, and prevented. Official identification of animals in national, intracommunity, and international commerce already takes place, while at the same time, identification of livestock that are vaccinated or tested under official disease control or eradication is also possible. Last, but not least, blood and tissue specimens can be accurately identified, and the health status of herds, regions, and countries can be certified (National Center for Animal Health Programs, USDA Web site).

8.2 Specific Requirements for Use of RFID Technologies in Animals

8.2.1 Means of Identification

The identification of agricultural animals in the United States was taking place for the first time at the end of the 19th and the beginning of the 20th century. Branding with a hotiron was used by cattle ranchers to indicate possession and to hopefully prevent cattle rustling. For keeping records, swine owners were employing ear notches. These two methods, however, stopped being popular due to their negative effects on production as well as for welfare reasons. Thus, the use of ear tags and tattoos appeared in the beginning of the 1960s (National Center for Animal health Programs, USDA Web site).

Dairy farmers have been keeping records for over 100 years. Each animal was identified and, in these records, was information mainly concerning reproduction performance. In 1950, there were several countries establishing identification systems as part of hygiene programs that were being mandated. This led actually to the establishing of two different identification systems with many disadvantages for the farmers, as this was less friendly, more expensive, and of lower quality (Wismans, 1999).

As it is obvious, many different means of identification have been used, such as ear tags, ear notches, tattoos, branding, paint marks, and leg bands. Electronic means of identification, which is comprised of barcodes and RFID transmitters (ruminal boluses, injectable transponders, and electronic ear tags) seem to have become more popular now for the traceability of animals (agricultural, domestic, and wildlife). In the European community, cattle are identified by ear tags. In the future, ear tags and RFID tags will comprise the most important means used in the identification and registration systems of agricultural animals.

8.2.2 Standards Regarding the Use of RFIDs: ISO 11784 and ISO 11785 and the Role of ICAR

As regards standardization concerning the use of RFID technology in animals, there are two international standards that apply: ISO (International Organization of Standards)11784 and ISO 11785. ISO 11784 refers to the Code Structure, while ISO 11785 refers to the Technical Concept.

According to the ISO standard, the carrier frequency for animal identification is 134.2 kHz, while there are three protocols in use to communicate between tag and reader. In regard to the bit stream used for communication between the transponder (RFID tag) and the transceiver (Reader), this includes an identification code, together with a code to ensure correct reception of the bit stream. This identification code is described in ISO 11784. On the other hand, use and handling of the transmitted information (activation of transponder, information transfer, protocols used) are described in ISO 11785. Therefore, the combination of the two above standards covers the use of RFID technology in the case of animals. In several cases, these standards are refered to as: ISO11784/11785. However, as the ISO, compliance to the above standards may involve the use of patents, especially regarding the transmission.

According to ISO 11784, a 64-bit identification code stored on transponders is used in which 10 bits are reserved for the country code, 38 bits are used for the main identification code within each country, and a 1-bit flag to indicate if an additional block of data, (user or case specific) is present. Following, there are 14 bits "reserved," and 1 bit to indicate an animal or nonanimal application use of the transponder. There is a specific numbering scheme for the identification of manufacturers, with codes 900 through 998 available for individual manufacturers and code 999 indicating a test transponder.

According to ISO 11785, five transponder types are described with respect to the protocols used:

- Full duplex system
- Half duplex system
- Destron technology system
- Datamars technology system
- Trovan technology system.

In practice, the first two are in full conformance to the standard. The full duplex system is used mainly for pet applications, while the half duplex system is used for livestock tracking and monitoring. The following table contains the specific technical parameters for each of the two cases.

System	Full Duplex (FDX)	Half Duplex (HDXS)
Operating Frequency	129–133.2 kHz/135.2–139.4 kHz	124.2 kHz = 1/134.2 kHz = 0
Channel Code	Differential Biphase	none
Symbol Time	0.23845 ms	0.1288 ms 1/0.1192 ms 0
Telegram (bit)	128	112

Though the ISO has provided the standards regarding aspects of RFID use in animals, it does not determine compliance of devices and tags with these standards. For this, based on a general agreement with ISO (Resolution ISO/TC 23/SC19N113, No. 45, August 1996), the International Committee for Animal Recording (ICAR) has developed procedures through which compliance of RFID systems can be verified with the standards. ICAR is an organization with more than 45 members worldwide focusing on standardization of procedures and methods for animal recording and testing for the approval of related recording equipment.

According to the ICAR-defined procedure for testing the compliance of RFID transponders with ISO standards, four types of ISO-compliant transponders for animal use are covered:

- Electronic ruminal bolus transponder
- Electronic ear tag transponders
- Injectable transponders
- Tag attachment

According to the test procedure, a manufacturer can apply for a full, limited, or listing upgrade test when requesting ICAR certification.

8.2.3 Evaluation of Current Electronic Means of Identification with Emphasis on RFID

It is widely accepted that identification and registration systems are necessary for animal traceability. National financial aid for systems for cattle, sheep, and goats demands reliable management, which depends on life-long identification of individual animals. Current methods are vastly in need of improvement, as they suffer from certain drawbacks, such as tag loss, breakage or alteration, slow data recording, and high error rates due to manual data transcription.

Passive electronic identification transponders consist of an electric resonance circuit, acting as a receiving/transmitting antenna, connected to an electronic microchip. When such a transponder is placed in an electromagnetic field of sufficient field strength, the induced voltage in the resonance circuit powers the microchip, which will turn on and start sending back its stored identification number via the resonance circuit. The reader, which generates the electromagnetic field, receives the transmitted code and displays it on a screen (Jansen and Eradus, 1999).

An electronic identifier device should remains on the animal for life and be recoverable after slaughter. In addition, it should be resistant to the hard farm conditions and provide information whenever needed. Finally, its cost should be low enough to permit its application to all agricultural animals.

A four-year program of the European Commission's Joint Research Center (Ispra, Italy) was implemented to encompass one million animals (cattle, sheep, and goats) and aimed at the evaluation of electronic means of identification. This IDEA program (Identification Electronic des Animaux) took place between 1998 and 2001 in six countries of the European community and revealed that improvement of identification and registration systems can be achieved via electronic means of identification.

In this program, three kinds of electronic means of identification were evaluated: electronic ear tags, ruminal boluses, and subcutaneous ear injectable capsules. Each of them contained an RFID transponder capable of transmitting the animal's identification code to a portable or static reader device.

The final report verifies that electronic means of identification can improve identification and registration systems. In addition, it points out that electronic ear tags can be applied in cattle, sheep, and goats of all ages, though ruminal boluses can be applied at a live weight of 25 kg for sheep and goats, and in cattle less than 20 days old. Finally, electronic ear tags and ruminal boluses present the lowest mean percentages of losses or damages (<0.5 percent) and, on average, their retention rates (>99.5 percent) are substantially higher than commonly observed retention rates for plastic ear tags (Institute for the Protection and Security of the Citizen Web site).

8.2.3.1 Ruminal Bolus

The ruminal bolus is a ceramic capsule retained in the reticulum or the second stomach of ruminants. It incorporates a radio frequency transponder capable of automatically communicating information to a reader. Ruminal boluses can be used as electronic means of ruminant identification.

There are different kinds of ruminal boluses available, for example, the ceramic, the monolithic, and the steel-weighted boluses. The ceramic boluses are cylindrical to enable oral administration in young animals. The ceramic material enclosing the electronic transponder is nontoxic. Small steel metallic objects near the capsule appear not to affect the reading distance of these transponders (Hasker and Bassingthwaite, 1996). The steel-weighted bolus was developed in The Netherlands by Nedap Agri and consists of the transmitter enclosed in glass, which is then enclosed in plastic. The steel-weight is positioned eccentrically.

Electronic passive devices incorporated in ruminal boluses have proven to be a safe choice for permanent ruminant identification (Caja et al., 1999). In the past few years, different kinds of boluses have been used in cattle, sheep, and goats without any negative effects on animals being reported. In spite of the size of the bolus, there doesn't seem to be any negative effects on the food intake, the digestion, or the morphology of the ruminal wall. However, there isn't a unanimous agreement on this among the scientific community. Some scientists imply that the transponder may alter, via mechanical action, the reticuloruminal mucosa and rumination patterns. Furthermore, the transponder may increase, via its electromagnetic action, the growth rate and metabolic activity of ruminal bacteria (Antonini et al., 2006).

At this point, it is appropriate to mention that foreign bodies can be introduced in the rumen without any negative effects on the ruminant's physiology. Magnets for metal foreign body entrapment and trace element, and antibiotic and growth promoter boluses are some examples.

In addition, in lactating ruminants, the development of their young animals' forestomachs can be helped by the existence of particles, as the forestomachs have not been fully developed in the first months. The development is particularly stimulated by providing solid food progressively, along with milk (Beharka et al., 1998). Foreign bodies in the forestomachs, as well as solid food for lactating animals, stimulate their motility and the rumination.

In recent studies, it is reported that the recommended live weight for ruminal bolus administration (66 × 20 mm, 65 g) in cattle, sheep, and goats is 30 kg, 25 kg and 20 kg, respectively. The retention rates for the boluses were 99.7, 100, and 98.8 percent, respectively. Caja et al. (1999) suggested that the use of the ceramic bolus is recommended as a safe and tamper-proof method for electronic identification of ruminants once the animals have reached a weight where successful administration is possible.

Other researchers suggest that boluses can be administrated in sheep long before they reach 25 kg. Twenty kilogram boluses can be administered to lambs after weaning at the fifth week, with weight of more than 12 kg. These kinds of boluses do not affect the age at the end of fattening or the mortality rate, and the retention rate is 100 percent (Garin et al., 2003).

8.2.3.2 Electronic Ear Tags

Electronic RFID transponders are embedded in ear tags, and they work as remotely activated receiver–transmitters, which use a short-range radio frequency. The ear tags are plastic-covered transponders capable of being fixed to the ear of an animal using a locking mechanism or to be attached in such a manner that it cannot be removed from the tag without damaging it.

The electronic transponders mounted in ear tags are an official, but not mandatory, option for identification of dairy cattle in North America and all cattle in the European Union, although visual tags must support the electronic identification in the European Union and in Canada. Externally mounted transponders are easily located and removed upon slaughter of the animal and are unlikely to enter the food chain. However, as any external tag is vulnerable to loss, removal, or damage, efforts to develop permanent electronic identification are focusing on internal forms, such as injected transponders (Stanford et al., 2001).

8.2.3.3 Injectable Capsules

The injectable capsules are also electronic means of identification, with electronic RFID transponders being implanted in several places in the animal's body. The transponders are covered by a capsule of biomedical glass, which is nontoxic for the animal. The preferable locations are the forehead, the external ear lobule, the posterior auricular base, and in the intraperitoneal cavity. In cattle, the optimum site of implantation is controversial. It has been proposed that the best site was in the base of the ear tissue, under the scutiform triangular cartilage (Fallon and Rogers, 1992; Fallon and Rogers, 1999). However, in a study on swine, the factors of skin damage and migration were analyzed, as well as the reading efficiency, and it was found that the best implant area for an injectable implant was the posterior ear base (Silva and Naas, 2006). The results of a study in adult sheep suggested that first the armpit, and second the ear base, as suitable subcutaneous injection sites for electronic identification (Caja et al., 1999).

The recovery of these electronic implants in abattoirs (slaughterhouses) is very important for two reasons. First of all, the recovery can prevent more than one use of the injectable capsule. Second, the recovery prevents the capsules from remaining in the food products with the risk of being consumed (Sheridan, 1991).

Injectable transponders meet the requirements of an identification system that is permanent, unique, tamper-proof, and does not produce apparent disturbances to the animals at application, especially for newborn piglets. The transponders can be applied shortly and easily after birth. The use of 23- to 34-mm transponders in the intraperitoneal position seems to be preferable (Caja et al., 2005).

8.2.3.4   Tag Attachment

These are transponder components covered by a primary protection layer and meant for producing one or more of the three transponder types mentioned above or other types of animal electronic transponders.

8.2.4   Comparison of the Electronic Means of Identification

Comparison of electronic and a simple means of identification has taken place in swine, suggesting that neither has any negative effect on animal health and performance. Farm losses averaged 1.6 percent for ear tags (visual, 0.8; half duplex, 1.9; full duplex, 2.7 percent; $P > 0.05$) and 1.8 percent for intraperitoneal-injected transponders (half duplex, 1.7; full duplex, 1.9 percent; $P > 0.05$). Moreover, 1.4 percent electronic failures occurred in the electronic ear tags (half duplex, 2.2; full duplex, 0.6 percent; $P < 0.05$), but not in the intraperitoneal-injected transponders. Visual ear tags and intraperitoneal-injected transponders were efficiently retained under conditions of commercial pig farms, which agrees with the minimum values recommended by the International Committee for Animal Recording (>98 percent), but when readability and reading ease were also included as decision criteria, injectable transponders were preferred (Babot et al., 2006). In addition, among the electronic ear tags and injectable transponders, another experiment suggests the second, (intraperitoneal implanted) as the best identification system for pigs (Caja et al., 2005).

The major advantage of passive injectable transponders is their small size, allowing an early application in young and small animals. One of the major disadvantages of the injectable capsules is the difficulty of recovery at slaughter; especially when the site is in the base of the ear tissue under the scutiform triangular cartilage, the recovery can be unpredictable. Due to the risk of entering in the food chain, there are researchers who consider the implantable transponders to be impractical (Fallon et al., Web site). On the contrary, the ruminal boluses can be easily recovered at slaughter by palpation of the reticulum, thus, possible contamination of food products with foreign bodies is avoided. However, the easiest electronic means of identification to be recovered at slaughter remains the electronic ear tag.

The Department of Environment, Food and Rural Affairs (DEFRA) published a preliminary study in 2003 on the cost of the implementation of the European Regulation No 21/2004 in the United Kingdom. This study concluded that the electronic means of identification was less costly than the conventional means. However, there is some disagreement in this field, particularly in Spain. According to Saa et al. (2005), the use of a mixed strategy combining conventional ear tags (for animals intended for slaughter) and electronic boluses (for breeding stock) seemed to be the most affordable strategy.

One of the major advantages of the ruminal boluses compared to the electronic ear tags and even to the injectable transponders is that they are more tamper proof. Their position makes it impossible for the bolus to be removed or replaced, whereas for the other two, replacement is feasible. In addition, the retention rates for boluses are higher than that of the electronic ear tags (Fallon et al., Web site).

8.3 Legislation

With the establishment of a single market in the European community, as well as the existence of both intracommunity and international trade, there arose new needs and demands for the improvement of the identification and registration systems. As a result, the first European regulation for national systems came into force in Directive 92/102/EEC, which made recording of animals and their movements obligatory. Before 1992, identification and registration systems regulations concerned only purebred animals (Wismans, 1999). Many regulations and directives have been enforced since then, with legislation for European countries nowadays being formed, as described below.

- Regulation (EC) No 1760/2000 recently came into effect, which establishes a system for the identification and registration of bovine animals as well as the labeling of beef and beef products. The most important points of this regulation, which repealed Council Regulation (EC) No 820/97 are: The system for the identification and registration shall comprise the following elements: ear tags, computerized databases, animal passports, and individual registers kept on each holding.
- All animals on a holding born after 31 December 1997 or intended for intracommunity trade shall be identified by an ear tag approved by a competent authority, applied to each ear. Both ear tags shall bear the same unique identification code, which makes it possible to identify each animal individually together with the holding on which it was born.

- After 31 December 1999, ear tags shall be applied within a period not longer than 20 days from the animal's birth. No animal may be moved from a holding unless it is identified.
- Any animal imported from a third (nonEU) country shall be identified on the holding of destination by an ear tag within a period not exceeding 20 days, and in any event before leaving the holding.
- The competent authority of the Member States shall set up a computer database, which shall become fully operational no later than 31 December 1999.
- From 1 January 1998, the competent authority shall issue a passport for each animal, within 14 days of the notification of its birth.
- Each keeper of animals shall maintain an up-to-date register; shall report within a period of seven days all animal movements, births, and deaths; and shall complete the animal's passports. The register shall be available at all times to the competent authority upon request.
- The Commission is examining, on the basis of work performed by the Joint Research Centre, the feasibility of using electronic means for the identification of animals.

Finally, in April of 2005, Commission Regulation (EC) No 644/2005 came into force, authorizing a special identification system for bovine animals kept for cultural and historical purposes on approved premises. According to this regulation:

- When the animals are moved to the premises, the approved ear tags may be removed from the animals without permission of the competent authority, but under his control.
- The animals shall be identified by one of the following means to be decided by the competent authority: two plastic or metallic ear tags, one plastic or metallic ear tag together with a brand marking, a tattoo, or an electronic identifier contained in a ruminal bolus or in the form of an injectable transponder.

Concerning ovine and caprine animals, Council Regulation No 21/2004 came into force in December of 2003, establishing a system for the identification and registration.

In 1998, the Commission launched a large-scale project on the electronic identification of animals (IDEA) and its final report was completed on 30 April 2002. That project demonstrated that a substantial improvement in ovine and caprine animal identification systems could be achieved by using electronic identifiers for these animals, provided that certain conditions concerning the accompanying measures were fulfilled. The technology for the electronic identification of ovine and caprine animals has been developed to the stage where it can be applied. Pending development of the implementing measures required for the proper introduction of the system of electronic identification communitywide, an efficient identification and registration system, enabling future developments in the field of implementation of electronic identification on a communitywide scale, should permit the individual identification of animals and their holding of birth.

- In accordance with the provisions of this regulation, each Member State shall establish a system for the identification and registration of ovine and caprine animals. The system shall comprise the following elements: means of identification to identify each animal, up-to-date registers kept on each holding, movement documents, and a central register or a computer database. The Commission and the competent authority of the Member State shall have access to all information covered by this regulation.
- All animals on a holding born after 9 July 2005 shall be identified within a period to be determined by the Member State as from the birth of the animal and in any case before the animal leaves the holding on which it was born. The period shall not be longer than six months. For animals kept in extensive or free-range farming sites, this may be extended to nine months.
- Animals shall be identified by a first means of identification: an ear tag. They shall be identified also by a second means of identification approved by the competent authority. The second means may consist in an ear tag, a tattoo (except for animals involved in intracommunity trade), a mark on the pastern (part of the hoof) (solely in the case of caprine animals), or an electronic transponder.
- Each keeper of animals shall maintain an up-to-date register. From 1 January 2008, and for each animal born after that date, the register must contain at least the following information: (1) the identification code of the animal, (2) the year of birth and date of identification, (3) the month and the year of death of the animal on the holding, and (4) the race and, if known, the genotype.
- From 9 July 2005, whenever an animal is moved within the national territory between two separate holdings, it shall be accompanied by a movement document.
- From 1 January 2008, electronic identification shall be obligatory for all animals. However, Member States in which the total number of ovine and caprine animals is 600,000 or less may make such electronic identification optional for animals not involved in intracommunity trade.
- From 9 July 2005, the competent authority of each Member State shall set up a computer database. From 1 January 2008, the database, with information for each separate movement of animals, shall be obligatory.

8.4 Systems Using Electronic Identification Devices and Electronic Data Transfer

During the past few years, different systems using electronic identification devices and electronic data transfer have been evaluated around the world. National authorities in many countries have commissioned pilot trials to evaluate the technology surrounding electronic means of identification and its systems, and where they can be applied.

The USDA has commissioned pilot projects and field trials concerning the National Animal Identification System (NAIS). These projects have very significant results with regards to NAIS implementation, since the projects support the findings that the electronic systems of identification and traceability can be implemented in real life. During 2004, 16 pilot projects took place in different States where low-frequency (LF) RFID technology was used. In one of these projects, the Florida Pilot Project specifically, (LF) RFID tags were used as a means of identification in 17,000 calves, 6,500 cull cows (adult cows with one or more calves and which are sold for slaughter), more than 12,000 calves and cows of the Seminole Tribe of Florida, and 200 horses. In this project, thousands of cattle were individually identified and the value of the electronic means of identification was demonstrated (USDA, May 2007).

In the United Kingdom, the Department of Environment, Food, and Rural Affairs (DEFRA) commissioned a pilot trial to evaluate the use of electronic means of identification and data transfer under English sheep farming conditions. Sixty-nine farms were used for the trial and over 122,000 electronic devices (HDX, FDX-B, tags, and boluses) were used. Electronic data could be transferred from a farm to the Project database via the Internet. From the conclusion, it is obvious that there should be improvement in the technology before instituting the application of the European Regulation (Final Report English Trial of Electronic Identification/Electronic Data Transfer in Sheep, 2005).

In these pilot trials, thousands of animals were identified by unique numbers and, thus, traceability was possible. At the same time, tag practical application, retention, and implementation were evaluated. Furthermore, the cost of RFID should be equated according to the species. Finally, further development in these systems is required to meet the farmer's needs (NAIS 2004 Initial pilot projects Final Report, USDA, May 2007; Final Report English Trial of Electronic Identification/Electronic Data Transfer in Sheep, 2005).

8.4.1 Databases: Opening Up a New Prospect

The electronic means of identification has been used in many different projects as part of traceability systems. One major advantage is that they can be used for organizing valuable databases on each farm; collecting data concerning identification, animal performances, and animal health. Such data can cover all aspects of an animal's life, performance, and health status, as can be seen from the prototype database example that is provided below. The database description is followed by a "per case" definition of methods for data collection.

8.4.1.1 Holding Animals Identification Parameters

A holding identification code is one of the basic elements that should be registered in every holding record. Identification parameters also include the holding address and the geographic coordinates or the equivalent geographic index of the holding. Animal species farmed and animal products gained from the farming activity also should be registered, together with the name and address of the holding owner. The last inventory record of animal stock and date of final removal of the herd should be registered. Total animal population per herd is dynamic information that should be registered once per year, while the three last successive inventory records should be kept in a list. Each animal should be identified and registered by its identification code consisting of two or three letters, corresponding to European Union (EU) country of origin (based on ISO 3166, Directive 21/2004 of EU Commission), and followed by its individual code. Individual codes consist of a maximum of 13 digits that form a unique number per individual animal of the particular country of origin. An individual code, combined with a country's code, form a unique identification code for each animal in the total EU animal stock. Date of birth and date of death occurring in the farm also are encoded as well as the date of tagging or insertion of identification device on the animal. The above information is necessary and, especially, in the case of sheep and goat farming where there are strictly defined time intervals wherein animals should be tagged and identified, otherwise penalties are imposed (Directive 21/2004 EU Commission).

Information concerning destruction, loss, subsequent replacement, or number of replacements of identification devices should always be registered. In all cases, an animal identification code remains the same, except for the case when the responsible authority permits the alteration of the code as long as traceability is not affected.

Dates of inspection and name of the official from the inspection authority should also be registered in the holding's records as well as all above listed data for the previous three years of the farming activity.

8.4.1.2 Animal Movement Parameters

Animal movement parameters are divided into two categories: input and output activity. Input relates to all new animals entering the farm. These animals are recorded based on their farm of origin's special code (which is unique per farm and obligatory) and supplemented by data related to the date of arrival, the number of incoming animals, and their identification codes. If animals are imported from a foreign country, they are given a new identification code, while, in the farm records, the primary identification code from the country of import is preserved and correlated to the new code.

Output activity notes all animals leaving the farm for the slaughterhouse. In this case, the following data is collected and registered: name of the contractor or merchant, number of the transport vehicles, identification code of the animals or name and address of the holding, code number or name and address of the slaughterhouse, and date of departure from the holding.

The above data is retained for at least three years.

8.4.1.3 Productive and Reproductive Data

In this field, the number of matings and artificial inseminations per conception is recorded. That is, a number is recorded in a list where n recordings, according to the animal species, are available. For example, numbers of the last four mating or artificial insemination attempts shall be kept in the list (2, 1, 1, 2), whereas every new number will delete the oldest recording. In addition, the number of gestations of every animal is recorded in the form of a serial number (1, 2, 3 ...). Moreover, the number of abortions is also recorded as a serial number as well as the kind and the number of pathological conditions of the genitals. The pathological conditions can be encoded in three to four letters and kept in a list of v recordings where every new one replaces the oldest (e.g., dystocia: DYS, metritis: ETR, etc.).

Concerning the reproductive performance, the number of neonatal and animals at weaning per sire is recorded in a list.

Also, data concerning milk production is recorded, including the duration of the lactic period and the quantity of milk from each animal during this period. This data can be listed in n recordings. The duration of the lactic period can be recorded as a number, which signifies days (e.g., 90, 145). Recording of the quantity of milk (per liter) can follow the same format.

Finally, animals' weight is recorded as a factor signifying growth and development. The weight is recorded in a list of numbers. The data kept per animal varies, depending on the species.

8.4.1.4 Animal Health Parameters Per Individual

Animal vaccination is a primary factor to be recorded. The commercial name of the vaccine (together with the name of the vaccine's manufacturer and distributor) and the administration date given as day–month–year should be registered. Furthermore, the past three year's vaccination programs should be recorded as well.

At the same time, all therapeutic and hormonal treatments are recorded by the commercial name of the product, the manufacturer and distributor company name, and the date of beginning and end of the treatment, encoded as day–month–year. Treatment background of the previous six months should be registered in a list, including the above information.

Finally, per individual animal, it is important to keep certain health data under the form of remarks or symptoms. It can be recorded (per animal) as:

1. Disease or symptoms, encoded by three to four letters, e.g., lameness: LAM, mastitis: MAST, etc.
2. Date of disease or symptom's occurrence and the date of recovery, given as day–month–year.

8.5 Recorded Parameters in Domestic and Wildlife Animals

Animals can be categorized according to their living conditions and surveillance needs. The purpose of this classification is to locate the parameters for recording of each animal in a specific category (Figure 8.1).

Following is a description of the parameters that need to be recorded and traced for each category.

8.5.1 Domestic Animals

8.5.1.1 Farm Animals

Identity—Identity characterizes each individual animal and is synonymous with unique codes consisting of certain letters and digits. Each animal's identity is unique and recognizable during its life and it is the primary element for the construction of animal identification and registration systems.

Heredity—This parameter determines the efficiency of all animal breeding systems. By the use of animal identification systems, desired performance traits are registered and observed through an animal's lifecycle. In this way, reliable databases are created and valuable information is available, especially for quantitative traits of high heredity co-efficiency, which helps the establishment of animal breeding programs at both the farm and national level.

Animal Tracking and Movement—Animal traceability is mainly secured with the establishment of identification systems. Farm or country origin is available at all times with a simple scanner reading and animal trade is better inspected at national, European, or international levels.

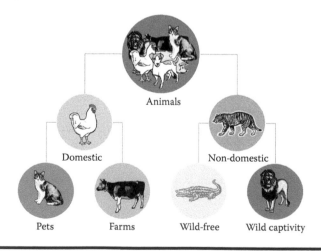

Figure 8.1 Animal classification for surveillance purposes.

Animal and Public Health—Based on animal identification systems, vaccination programs, infectious disease eradication programs, and, in general, herd hygiene management are easier to follow. On the other hand, animal traceability is further linked to animal products traceability, which ensures better inspection all the way from the farm to the supermarket shelf. In this way, products of animal origin that do not follow the hygiene standards and are potentially harmful to the public's health, are immediately detected and withdrawn from human consumption.

8.5.1.2 Companion Animals

Identity—(See Farm Animals above).

Animal Tracking and Movement—Animal traceability is of major importance in companion animals as well, as it ensures the recovery of lost or kidnapped animals.

Animal and Public Health—Based on identification systems, vaccination programs for companion animals (especially for zooanthroponoses like rabies), infectious disease eradication programs, and quarantine measures are easier to follow.

8.5.2 Nondomestic Animals

8.5.2.1 Wild Animals

Surveillance and Study of Behavior—By using electronic tagging systems or similar updated identification systems, gathering information and studying the behavior of wild animals, birds, fish, or even insects living in nature is becoming very safe and efficient. Information concerning wild animals' mobility and migration, nutritional behavior, and location of reproduction activity become significant especially for rare or endangered species whose survival most likely lies in human hands (Block et al., 2003; Streit et al., 2003).

Animal Tracking and Movement—Traceability helps prevent the illegal trade activity of wild animals and ensures endangered species defense as well as public health protection, by blocking the entry of zooanthroponotic exotic diseases (e.g., influenza, rabies, etc.) to disease-free countries.

Rehabilitation Back to Nature—Finally, rehabilitation of captured or injured animals back to their natural environment is better surveyed by using identification systems. These systems give not only the opportunity to record and study animals' behavior when inserted back into the wild, but also help the experts to drastically interfere in case of an accident or when rehabilitation programs fail.

8.5.2.2 Animals in Captivity

Identity—In captured animals, identity (animal passport) has the same significance as in domestic animals, by adding the extra benefit to control in zoo and animal parks for the legal possession of these animals (e.g., CITES—Convention on International Trade in Endangered Species of Wild Fauna and Flora program).

Animal Tracking and Movement—Traceability is also important especially when park-to-park movements take place and animal trade is made between circuses, zoos, and animal parks. In all cases, fast and efficient inspection of all captured animals' movements is successful when animals are electronically tagged.

Interactive Guide Systems—Electronic tagging also can be applied in guide systems for animal parks and zoo visitors. Specifically, these systems rely on RFID transponders that are placed in the vicinity of the corresponding animal and visitors, through which a visitor's scanner device can access information about each species (Hlavacs et al., 2005).

8.6 Open Issues

Though the use of RFID technology in animal identification, tracking, and monitoring has been a practice for a long period of time, there are still issues that have to be resolved. One very important aspect is that of a global standard for information recording regarding all categories of animals. ICAR has released a document (ICAR, 2001) regarding the parameters that should be traced in beef. However, a global standard covering all types of animals and establishing a common ground on which information regarding animal traceability can be collected does not exist. Such a standard should take into consideration:

■ The ability to track information globally and across borders
■ Compatibility with all types of RFID tag technologies, according to the animal case
■ Certification and security issues (antitampering, replication of information, etc.)
■ Links and interfaces with the food-tracing systems for cases where animals or their products are used for the production of food (e.g., meat, milk).

Such standards should go beyond the basic framework defined by the ISO on one hand, and expand the limited case studies presented by ICAR on the other.

Another important aspect is that of security. In the past few years, concerns about the use of RFID and the possible risks that may be hiding behind its use, as regards electronic attacks, even up to the case of RFID-based viruses (Rieback et al., 2006) have been expressed. In one of the real-world scenarios described in the RFID viruses and worms Web site (www.rfidvirus.org/), the case of a prankster that launches a RFID virus attack using commercially available equipment is described. According to the scenario, the attacker introduces a virus in the RFID tag containing the ID of his pet cat and then goes to a veterinarian (or the ASPCA), claiming that it is a stray cat and asks for a cat scan. Through this scan, the database containing the data of animals is infected, and, subsequently, when new tags for animals are produced, these are also infected and can also infect all databases in systems where they are scanned in the future. This creates a new form of virus that is transferred from animals to computer systems and vice versa, unlike traditional animal or even computer (if the term traditional can be used here) viruses. Such a scenario unveils a horrifying case where an attack to a government or widely used system may be launched resulting in the loss of millions of dollars. For this, scientists and technicians who are working in the field of RFIDs, warn that it is now, with such systems still in their infancy, that we should act fast and secure them with the necessary mechanisms as well as methodologies for the design and implementations that incorporate security (both in the technology used and the supporting software platforms).

Finally, an issue that offers a great field of research is that of the part of the animal body where the injectable capsule should be implanted. As mentioned above, different opinions have been expressed and the experience and research results so far indicate that there is no definite or unique answer here. Instead, the type of animal and also the framework in which the RFID chip is going to be used, should be taken under consideration for determining the place of injection. Therefore, a guideline following research results for the best placement of RFID chips, together with guidelines for selection of the best format of RFID chips, is needed, and offers an opportunity for new research work in the field of animal RFID tagging.

References

Antonini, C., Trabalza-Marinucci, M., Franceschini, R., Mughetti, L., Acuti, G., Faba, A., Asdrubali, G., and Boiti, C. (2006) *In vivo* mechanical and *in vitro* electromagnetic side-effects of a ruminal transponder in cattle, *J. Anim. Sci.*, 84: 3133–3142.

Babot, D., Hernandez-Jover, M., Caja, G., Santamarina, C., and Ghirardi, J.J. (2006) Comparison of visual and electronic identification devices in pigs: On-farm performances, *J. Anim Sci.*, 84: 2575–2581.

Beharka, A.A., Nagaraja, T.G., Morrill, J.L., Kennedy, G.A., and Klemm, R.D. (1998) Effects of form of the diet on anatomical, microbial, and fermentative development of the rumen of neonatal calves, *J. Diary Sci.*, 81: 1946–1955.

Block, B.A., Costa, D.P., Boehlert, G.W., and Kochevar, R.E. (2003) Revealing pelagic habitat use: The tagging of Pacific pelagic program, *Oceanological Acta* 25: 255–266.

Caja, G., Ribo, O., and Nehring, R. (1998) Evaluation of migratory distance of passive transponders injected in different body sites of adult sheep for electronic identification, *Livestock Prod. Sci.*, 55: 279–289.

Caja, G., Conill, C., Nehring, R., and Ribo, O. (1999) Development of a ceramic bolus for the permanent electronic identification of sheep, goat and cattle, *Comp. Elec. Agric.*, 24(1): 45–63(19).

Caja, G., Hernandez-Jover, M., Conill, C., Garin, D., Alabern, X., Farriol, B., Ghirardi, J. (2005) Use of ear tags and injectable transponders for the identification and traceability of pigs from birth to the end of the slaughter line, *J. Anim. Sci.*, 83: 2215–2224.

DEFRA Final Report English Trial of Electronic Identification/Electronic Data Transfer in Sheep, 2005, http://www.defra.gov.uk/animalh/id-move/sheep-goats/eid/pdf/ept-final.pdf, accessed May 2007.

Fallon, R.J., and Rogers, P.A.M. (1992) Use and recovery of different implantable electronic transponders in beef cattle and calves. *Irish J. Agri. Food Res.*, 31: 100–101.

Fallon, R.J., and Rogers, P.A.M. (1999) Evaluation of implantable electronic identification systems for cattle, *Irish J. Agri. Food Res.*, 38: 189–199.

Fallon, R.J., Rogers, P.A.M., and Earley, B., "Electronic Animal Identification, ARMIS No.4623, Irish Agriculture and Food Development Authority." http://www.teagasc.ie/research/reports/beef/4623/eopr-4623.htm, accessed May 2007.

Garin, D., Caja, G., and Bocquier, F. (2003) Effect of small ruminal boluses used for electronic identification for lambs on the growth and the development of the reticulorumen, *J. Anim. Sci.*, 81: 879–884.

Hasker, P.J.S., and Bassingthwaite, J. (1996) Evaluation of electronic identification transponders implanted in the rumen of cattle, *Aust. J. Exper. Agri.*, 36(1): 19–22.

Hlavacs, H., Gelies, F., Blossey, D., and Klein, B. (2005) A ubiquitous and interactive zoo guide system, *INTETAIN, LNAI* 3814: 235–239.

ICAR (July 2001) Beef Recording Guidelines: a synthesis of an ICAR survey. ISSN: 1563–2504. ICAR Technical Series 6, accessed May 2007.

Institute for the Protection and Security of the Citizen. http://ipsc.jrc.cec.eu.int, accessed May 2007.

Jansen, M.B. and Eradus, W. (1999) Future developments on devices for animal radio frequency identification, *Comp. Elec. Agri.*, 24: 109–117.

McKean, J.D. (2001) The importance of traceability for public health and consumer protection, *Rev. Sci. Tech. Off. Epiz.*, 20(2): 363–371.

Rieback, M.R., Crispo, B., and Tanenbaum, A.S. (March 2006) Is your cat infected with a computer virus,? *Fourth IEEE International Conference on Pervasive Computing and Communications (PerCom'06)*, Pisa, Italy, pp. 169–179.

Saa, C., Milan, M.J., Caja, G., and Ghirardi, J.J. (2005) Cost evaluation of the use of conventional and electronic identification and registration systems for the national sheep and goat populations in Spain, *J. Anim. Sci.*, 83: 1215–1225.

Sheridan, M.K. (1991) Electronic identification systems: governmental considerations. In *Automatic Electronic Identification Systems for Farm Animals*. Proceedings of a seminar held in Brussels, Belgium, E. Lambooij (Ed.), 13198: 109–113.

Silva, K.O. and Naas, I. (2006) Evaluating the use of electronic identification in swine, *Eng. Agríc.*, 26(1): 11–19.

Stanford, K., Stitt, J., Kellar, J.A., and McAllister, T.A. (2001) Traceability in cattle and small ruminants in Canada, *Rev. Sci. Tech. Off. Int. Epiz.*, 20: 510–522.

Streit, S., Bock, F., Pirk, C.W.W., and Tautz, J. (2003) Automatic life-long monitoring of individual insect behavior now possible, *Zoology*, 106: 169–171.

USDA, "National Center for Animal Health Programs,." http://www.aphis.usda.gov/vs/nahps/animal_id/, accessed May 2007.

USDA (May 2007) NAIS 2004 "Initial Pilot Projects Final Report." http://www.aphis.usda.gov/newsroom/content/2007/05/content/printable/PilotProjectReportFINAL05-01-2007.pdf, accessed May 2007.

Wismans, W.M.G. (1999) Identification and registration of animals in the European Union. *Comp. Elec. Agri.*, 24: 99–108.

Chapter 9

RFID Applications in Assets and Vehicles Tracking

Wei Liu, Zhao Peng, Wenqing Cheng,
Jianhua He, and Yan Zhang

Contents

9.1 Introduction

RFID (Radio Frequency IDentification) is a contactless interrogation method of automatic wireless identification and data acquisition. A RFID system essentially consists of three components: RFID tag, RFID reader, and backend information system. The RFID tag has a readable and writable memory chip together with an antenna, which can be attached or incorporated into a product, animal, and so on. The RFID reader device can communicate with multiple RFID tags simultaneously via radio frequency waves. The backend information system can provide the corresponding information of products attached by RFID tags. In the proposal of EPCglobal, RFID information can be shared via EPCglobal network globally [1][2].

Compared to barcodes, there are some evident advantages of RFID: (1) contactless and remote interrogation, (2) no line of sight required, (3) multiple parallel reads possible, and (4) individual items instead of a specific class of items can be identified [3]. As a result, there could be a dramatic improvement in process and operation by using RFID in object identification. Every object can be tracked and the detail information for individual objects can be provided with minimal operation delay. In fact, the usage of RFID is not just limited in the data collection process, it will bring great benefit when integrated to existing information systems. A number of companies provide RFID technology solutions for application integration, such as SAP®, Oracle®, IBM®, Microsoft®, and Sun®. These solutions make RFID more feasible and practical to be widely deployed. Large retailers such as Wal-Mart, Target, and Albertsons have begun to deploy RFID systems in their warehouses and distribution centers, and they require their suppliers to add RFID tags on product pallets [4].

Since the technical challenges are different in different RFID scenarios, here we take a certain steel manufacture enterprise in China as the case and discuss the technical issues in its asset and vehicle-tracking application. This steel manufacturer consists of a number of factories located nearby. Raw and processed materials are transported among warehouses and plants via vehicles. Each year, there are millions of Chinese yuan lost due to the incorrect delivery or stolen during intra-transportation. It is difficult to track the logistic flow in such a big enterprise with traditional methods. Then RFID technologies are introduced to track the assets and vehicles involved in this enterprise. RFID tags are installed on thousands of transport vehicles and attached to the key industry materials. RFID readers are set up at the main gates and at the warehouses. When transport vehicles pass through the gates, the RFID tags' ID of vehicles and goods are acquired by the RFID reader. The data is uploaded to the Information Monitoring Center and checked, and then the verification information is fed back to the safeguard at the main gates. Since the enterprise is spread over such a relatively large area, it is required that the RFID tags be read and checked at temporary checkpoints. Also, the transportation information of current goods in the enterprise resource planning (ERP) system is also needed to check the vehicles.

In this RFID application, the mature RFID solution in ultrahigh frequency (900 to 915 MHz) for nonstop vehicle management was adopted. According to the above application requirements, simply deploying RFID tags and readers is not enough. At first, asset query function is expected in any place at any time, which calls for the integration between RFID and wireless communication. This will enhance the mobility and portability in actual enterprise RFID application, and ensure the real-time tracking on vehicles and asserts [5][6]. Second, seamless integration with existing information systems is also expected. As a result, RFID middleware is needed, which cannot only support different kinds of RFID readers hardware [7][8], but also support integration to existing enterprise systems, such as ERP, supply chain management and SCM (supply chain management). This will also increase the reliability and scalability of RFID applications. As a result, Wireless Intelligent Terminal and Smart RFID Middleware systems have been developed and adopted.

The organization of this chapter is as follows. Section 9.2 introduces the application scenario and problems for RFID-based tracking, and briefly describes the main technical challenges and developing tasks. Section 9.3 states the design and implementation issues of the Smart RFID Middleware system and the Wireless Intelligent Terminal. Section 9.4 describes the RFID application in two major systems: warehouse management and vehicle management systems. Finally section 9.5 summarizes the chapter and provides some discussion on open issues.

9.2 Problems

9.2.1 The Model for RFID-Based Tracking

Tracking application is very common in an asset security and logistics system. With the help of tracking, enterprise can instantly handle the asset location, operation status, and goods circulation information, so as to realize strict control of inventory and the avoidance of low turnover ratio. For the convenience of this discussion, we use a logistics delivery process to discuss a tracking model.

As shown in Figure 9.1, most logistics systems can be simplified into a three-party model, which involves the role of seller, transporter, and buyer. We can also extend it to a general asset delivery case just by replacing the seller and buyer with sender and receiver. In the actual process, the buyer first sends orders to the seller; after receiving the orders, the seller sends the goods and submits a goods list to the transporter; when the buyer finally receives the goods, it returns acknowledgement information to the seller. Figure 9.1 shows the detailed five-step information flow interaction between the three parties.

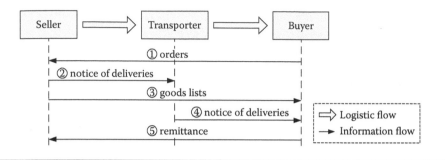

Figure 9.1   Tracking model without RFID.

In traditional tracking cases, the goods can be identified by barcode. However, it is well known that barcoding has some shortages, such as the sight identification, single-item reading, and manual operation, which significantly reduces the efficiency of the tracking process. Compared to barcode, RFID can reduce the circulation time and manual errors, with the characteristics of contactless and remote interrogation, multiple-parallel readings. Then RFID technique can realize automatic information acquisition, which can reduce the time delay among the information flow interactions.

Figure 9.2 shows the benefit of RFID deployment in the tracking process. The RFID readers are installed at the gateway position, e.g., the door of the warehouse. When the goods reach each party, a RFID device will read the goods, and RFID middleware can find out the corresponding information of the goods, such as quantity, characteristics, etc. Then the notice of deliveries or remittee information can be automatically delivered by the software triggers in the RFID middleware. What's more, RFID middleware can report the RFID events to a monitoring center, and then the goods location and operation status can be tracked in real time.

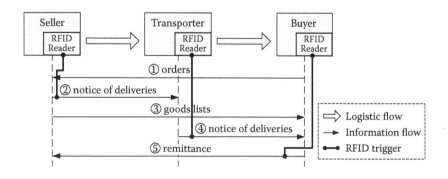

Figure 9.2   Tracking model with RFID.

Thus, the logistics and information flow in this model can achieve ideal synchronization resulting in increased efficiency of whole logistics delivery process.

If we extend this three-party model into an actual asset and vehicle tracking scenario, two issues should be emphasized. First, the three parties can be extended to an unknown number of parties in the tracking chain, including some temporary parties, e.g., a temporary monitoring station along the road for vehicle tracking. In that case, wireless networking technologies are needed to enable ubiquitous access to RFID information on the Internet or Intranet. Second, a large number of RFID events should be filtered in time and trigger proper actions. Then, RFID middleware is needed to handle this RFID data and to integrate it to existing enterprise backend information systems. In the following section, we will discuss in detail these two technical challenges.

9.2.2 Technical Challenges

9.2.2.1 How to Achieve Mobility and Portability Requirements for RFID Applications

In an asset and vehicle tracking application, it is required to support the deployment of dynamic tracking. Since the fixed RFID reader can no longer meet movable and portable requirements, a handheld style of RFID reader is then needed. On the other hand, in order to access the RFID information on the Internet or Intranet, this handheld reader should support not only wireless networking, but also have the ability to process, e.g., providing GUI (graphical user interface) for RFID data query and verification. In a word, this handheld equipment should support RFID data acquisition, RFID information process, and wireless network access. In this chapter, it is called wireless intelligent terminal (WIT).

In the temporary monitoring station, the guarders can use WIT to acquire RFID tag data from vehicles and goods, and then query monitoring centers via wireless access technology to obtain the vehicle information and logistics information. They can also use WIT to input or update the transportation status information attached to RFID tags. The ability of wireless network access brings great convenience to dynamic asset and vehicle tracking. In some cases, the monitoring center in the backend system can even send operation or scheduling orders to terminal holders.

In our development, we extended an existing embedded information process terminal to support wireless network access and the RFID reader. However, there are two technical challenges:

■ Hardware integration of wireless communication supporting multiple wireless access modes, such as dial-up wireless access via general packet radio service (GPRS) and code division multiple access (CDMA), IEEE 802.11 wireless local network, etc.
■ Software driver to configure and control the wireless network interface and plug-in RFID reader modules.

9.2.2.2   How to Seamlessly Integrate an Enterprise System

In real deployments, there will be a number of RFID readers or devices at different places. How to manage and configure these devices is one key problem that needs consideration. Another issue is how to transfer large amounts of raw RFID tag data between RFID devices and backend systems. RFID middleware is the immediate software to tackle these two problems. Furthermore, some extension functions, such as event filtering, are also important to reduce the data rate to backend information systems.

For instance, in an inventory management scenario, people usually configure all the RFID readers or devices in the initial phase. Configuration includes the reading range, the type of tag, and the event filtering rules. All that work can be done by RFID middleware. When goods with RFID tags enter into the reading range of a RFID reader, the RFID reader only reads the tags' raw data. RFID middleware can query tags' information from the basic RFID service. Then it can call the pre-defined procedures (such as inbound and outbound) in the backend system, or control some physical devices (such as green/red alert light). These kind of software functions are called RFID triggers in this chapter.

There are three major technical challenges in RFID middleware development:

- Device management: To configure or control connected RFID readers or devices.
- Event filtering function: To set various event rules, such as inbound, outbound, pass, etc. and filter the useful data.
- Interface management: To manage various software interfaces between RFID middleware and the backend system, such as Web Service, CORBA, SOAP, etc.
- Application integration: to integrate RFID middleware with the backend system via various software interfaces, such as JMS, SOAP, etc.

9.3 Technical Solution

9.3.1 Wireless Intelligent Terminal

Nowdays, RFID terminals with wireless communication can make goods identification much more convenient. A user can use a handheld RFID terminal to obtain the tag data and then use a kind of wireless communication to get the additional user-trackable information of the objects [9].

It is quite common that the companies provide their proprietary roaming protocols to work on distribution systems with WLAN (Wireless Local Area Network). The natures of WLAN are easy to build, flexible, using open industrial, scientific, and medical (ISM) frequency, and able to support mobile station communication [10]. It has the irreplaceable advantage of being able to maintain a high-speed mobile communication under constraints, such as lack of communication resources and wiring problems related to the physical environment.

In addition, the WWAN (Wireless Wide Area Network) technology, such as GPRS and CDMA, can also be applied in RFID terminael because it provides dial-up wireless network access that the wireless terminal can use when there is no WLAN. With the WWAN, the terminal outside can access the Internet[11] and communicate with a lower speed than WLAN.

The RFID technology should co-exist with wireless technologies to transmit the real-time data[12]. The WIT is special terminal equipment based on the embedded information process device. It combines data acquisition, data processing, and mobile data communication functions together. The WIT can fulfill the frequent RFID or barcode data acquisition task, and, meanwhile, it also can provide real-time communication to the data/or WLAN module center using the CDMA/GPRS wireless LAN module.

9.3.1.1 System Framework of WIT

The WIT is mainly composed by terminal equipment and external modules. The terminal equipment provides a general computing platform with an embedded TFT screen and input module, and it can embed different kinds of wireless communication modules according to the users' requirements. WIT supports voice communications, mobile computing, and wireless accessing at one time. It also provides a variety of standard interfaces for different kinds of special logistics equipments, including barcode scanners and RFID reader modules. The components of WIT are shown in Figure 9.3.

9.3.1.2 Hardware Design of WIT

The hardware design work of WIT includes the choice of a microprocessor or platform. A proper embedded microprocessor is chosen depended on the requirements of the real application. Then a broad platform is chosen to achieve smaller size and more integrated function of the hardware system. In order to support external hardware, such as plug-in RFID readers and wireless communication modules, this

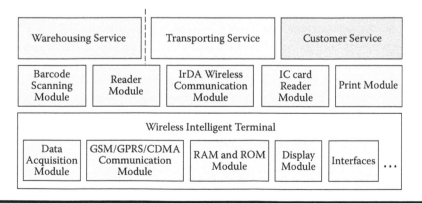

Figure 9.3   System framework of WIT.

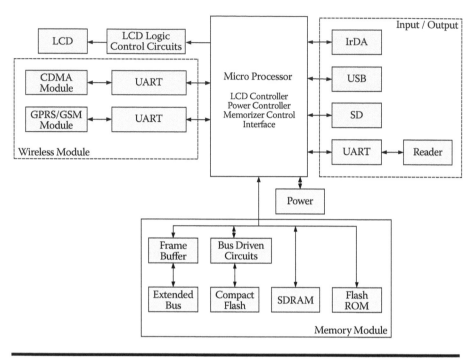

Figure 9.4   Hardware architecture of WIT.

platform should provide different kinds of standard interfaces with a united power supply. The hardware architecture of WIT is shown in Figure 9.4.

As shown in Figure 9.4, this hardware system consists of microprocessor module, memorizer module, wireless communication module, display module, I/O module, etc. Among these modules is a specially designed wireless communication module. The communication modes of wireless WAN are dial-up wireless access of GPRS or CDMA in China. GPRS has low data rate while been widely used; CDMA has higher data rate, but smaller network coverage than GPRS in China. Here, we supports both modes. At present, WIT mainly uses the infrared and wireless LAN card to implement wireless communication. The infrared interface uses the interface of a central processing unit (CPU), and the wireless LAN card uses a PCMCIA card, which is very easy to use in the embedded system.

9.3.1.3   Software Design of WIT

The embedded software design work of WIT includes information process modules, embedded RFID middleware modules, and wireless network management modules. Information process modules provide GUI interface for the user to input and read information. Embedded RFID middleware provides drivers for the RFID reader module and deals with the RFID tag data. Wireless network management modules provide drivers to wireless communication modules, manages the wireless network connection, and supports SMS (Short Message Service) functions.

As we know, SMS service is very popular in China and has been widely used as a kind of information delivery method. As a part of enterprise requirements, the acquired RFID tag data could be reported to a specific user by a SMS message service. This kind of communication method can be a candidate communication method to realize information exchange between machines or between human and machines when there is no Internet connection. So, the wireless network management module provides API (application programming interface) to send and receive SMS messages. On the other hand, because the length of a SMS message is limited, embedded RFID middleware is required to assemble original RFID data into multiple readable SMS messages. It should be noted that, SMS service is not a reliable communication method, which means the SMS message could be lost when a wireless base station is busy. As a result, the SMS interaction is not recommended to exchange information between machines, but can be applied in the monomial information transmitting (noticing the status information) from machine to humans.

9.3.2 Smart RFID Middleware Systems

RFID middleware is a kind of software that lies between the frontend of RFID reader module and the backend of the enterprise database or software application system (such as ERP, CRM (customer relationship management), and WMS (warehouse management system)). We can also regard it as the platform to manage and route the RFID data from the RFID reader device to the enterprises application system. The deployment of RFID middleware is shown as Figure 9.5.

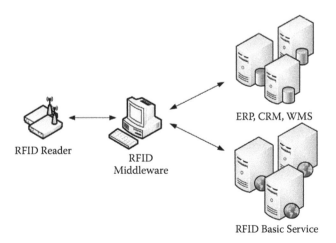

ERP, CRM, WMS

RFID Reader

RFID Middleware

RFID Basic Service

Figure 9.5 Deployment of RFID middleware.

RFID middleware simplifies and clarifies the interface of the RFID data source to the application system of the enterprises by monitoring precisely the bottom-layer devices, real-time collection of the original data as well as data filtering and encapsulating typical application logic. Additional functions include the sustaining of readers, device management, application integration, flowing management, etc. In order to realize the seamless data connection and application integration from the readers and the application system of the enterprises, the RFID middleware needs to interact with the application system and the RFID basic service system, such as ONS (object naming service) and EPC-IS (EPC information service) servers in EPCglobal architecture.

9.3.2.1 System Architecture of Smart RFID Middleware

The configuration of the RFID middleware system that supports the multicommunication platform is shown in Figure 9.6. It consists mainly of five functional modules: RFID device management, event management, data management, network communication management, and interface integrating management.

The functions of the five modules are as follows:

- RFID device management module is in charge of connection and control RFID readers from different manufacturers. It provides the unified management interface to various RFID reader devices with common control commands set.
- Event management module uses event filters to obtain effective RFID events from original raw RFID events, and transmit them into the data management module.
- Data management module deals with the effective events generated by the event manager. It extracts useful information and stores it in a local database in the RFID middleware.
- Network communication management module manages the protocols of network communication and maintains the data transmitting queue. The middleware would shield the distinction of the approaches of the bottom-layer communication, and it would provide a unified mutual channel of information for a lot of RFID readers and enterprise applications.
- Integrate interface management module provides different application integration or data exchange method connecting backend system, as well as the public RFID basic service, such as EPCIS.

9.3.2.2 Event Filtering

Event filtering is an essential function of RFID middleware. Because the original RFID events obtained by the reader only describe the fact that the tag is in the sensing area, they are far from the event level defined by the backend information system.

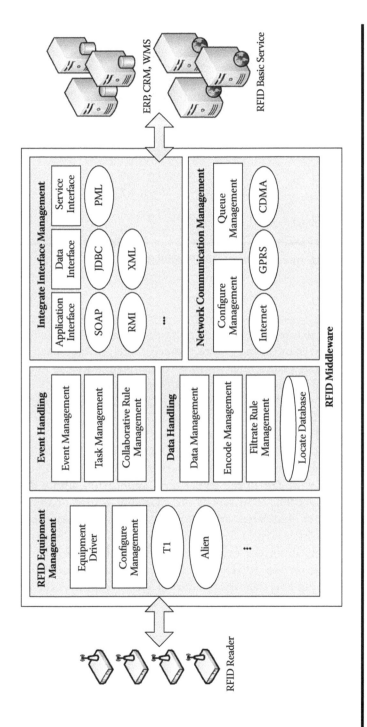

Figure 9.6   System architecture of smart RFID middleware.

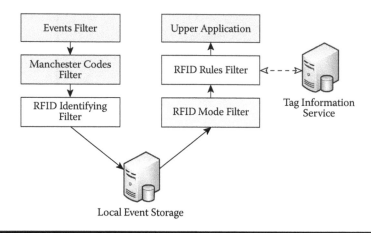

Figure 9.7   Filter management of Smart RFID middleware.

Filtering management needs to set several filtering rules. Smart RFID middleware realizes filters on five levels, including an effective event filter, a Manchester codes filter, an identifying filter, a rules filter, and a formula filter. All these filters are cooperated one by one. The original raw RFID events from the RFID readers will be filtered into effective RFID events, which can be accepted by backend enterprise system. A local event storage database is used to store temporary RFID required by the filters. The process is shown in Figure 9.7.

The tag event that the RFID reader identifies is the main data source of the RFID filter. There are two ways to describe a RFID event—one is to report the condition of the tag to the upper layer application with a settled period of time, and the other is to report the changing course of the event to the management layer. The latter one is similar to the Manchester coding in the circuit design, which largely decreases the information redundancy in the event queue by describing the existing status of the label. So, adopting the Manchester coding on the RFID event is an essential precondition for the speediness of filtering and the effectiveness of identification of the RFID event. Meanwhile, filtering the RFID event queue can eliminate the influence brought by the unsteadiness while reading the signal. After the events are made into sequences, we can define the mode according to different initial state and transition processes, and the filter will attach the timing information to each single unit of the mode sequence.

In most circumstances, a RFID reader needs to deploy several filters in order to meet different application needs. For example, in the future, a shelf reader deployed in a supermarket may face the identification of several event modes at the same time: (1) adding mode, i.e., complementing the goods to be sold on the shelf; (2) departing mode, i.e., the consumer taking goods away from the shelf; (3) resting mode, i.e., the consumer might be picking up the goods and his/her shopping cart may have rested next to the shelf for some time; and (4) passing mode, i.e., the consumer passing the shelf with his/her shopping cart.

All four fundamental identification modes of RFID events need to be supported simultaneously, thus, a process that merges the description scripts of several filters is necessary. A Smart RFID middleware system implements the merging of several filters by applying multiway trees. According to the merged combination of filters, a complete identification mode tree can be generated in the internal memory. Through the filtering of the identification tree, all the event queues would stay in a stable transition state, then, the steady latency of the system would end the identification by force. Thus, the final coding status is gained. If we code the path that the final state has passed through in each layer of the tree, the eventual state of coding is corresponding to the digital coding of the filter. Since our RFID middleware can support many data filter modes and their extension modes, we call it "Smart" RFID middleware.

9.4 Application Cases

In this section, we discuss the application case of a certain Chinese enterprise. The enterprise needs real-time information to make the decision of logistics management, and the area of factory is too large to manage, so there are big security loopholes in the area of artificial inspection. Furthermore, the lack of accurate and timely vehicle information often leads to theft resulting in unnecessary economic losses. So, it is necessary to track and record the information of vehicles.

As the introduction mentioned, using a wireless intelligent terminal and Smart RFID middleware system could achieve excellent vehicle and asset tracking applications. According to the different objects to be identified (such as goods or transportation), it can use different frequency RFID tags. In this enterprise, we attach the high frequency (13.56 MHz) or ultra-high frequency (900 to 915 MHz) tags to the goods according to its composition and cost, and fix the ultra-high frequency tags on the vehicles to support nonstop checking. Then RFID readers are placed on the doors, conveyer belts, the storage shelves, and pick-up areas in the warehouse to collect the information of the high frequency tags on the goods. Also the ultra-high frequency RFID readers are placed on gates and important roads.

For the requirements of mobile and dynamic asset tracking, we develop a kind of wireless intelligent terminal to read RFID tags and obtain related information from monitoring centers via wireless communication. The wireless module can be wireless LAN, if it exists, or sent by CDMA dial-up connection. Smart RFID middleware is also deployed to provide scalable analysis and detection of the real-time RFID information. The middleware is also integrated to the backend information system with various software integration interfaces.

RFID-based logistics management systems increase transparency of logistics management and automatization of inventory management, so that goods and transport vehicles can be tracked real-time to get their clear position, number, and identity. In the RFID-based warehouse management application, goods report management can be completed simultaneously in the progress of inbound or outbound. In the RFID-based vehicle tracking application, the attached RFID tag can report the logistical information of the vehicles. Two main application scenarios will be introduced in detail below.

9.4.1 Warehouse Management

Typical warehouse management application provides the functions of inventory management, inventory change, material requisition/returning record, etc. It is important to provide various inventory status reports instantly to managers. Warehouse management systems involve system management, storage/retrieval operation management, order form management, plan for delivering of goods, report management, inquiry management, sales return management, data maintenance, etc. Deployment of RFID in the warehouse management system will change the internal data flow. RFID interface systems are integrated with various software modules, as shown in Figure 9.8.

RFID-based storage/retrieval examination modules provide functions of inbound or outbound checking and location tracking management. In the traditional barcode-based method, the handset terminals scan the barcodes attached to goods and report the collected data to the backend server. Due to lower scanning efficiency and less flexibility of barcodes, there are some strict limitations with the machine-to-machine automatic information exchange. The process of a RFID-based solution is shown in Figure 9.9. RFID readers automatically collect all goods information while they are passing through the gate of a warehouse, and then complete counting management and report the goods information to the database of warehouse management system.

After goods have been inbound, the RFID-based tracking management system will handle the delivery process of goods, as shown in Figure 9.10. Fixed RFID readers automatically count and check the goods, then the inventory information is updated. In the process of picking, circulating, and processing and packing, goods can be tracked by the RFID readers distributed throughout the warehouse.

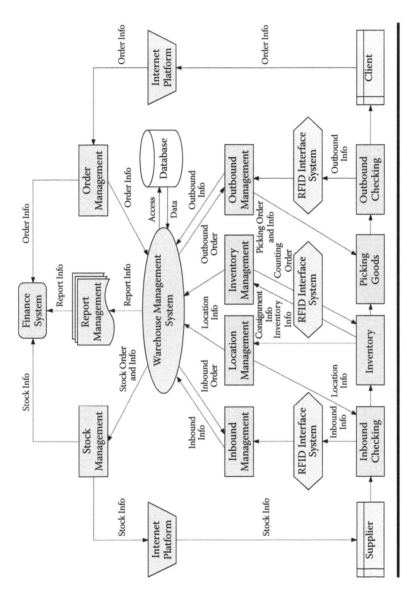

Figure 9.8 Warehouse management system dataflow process.

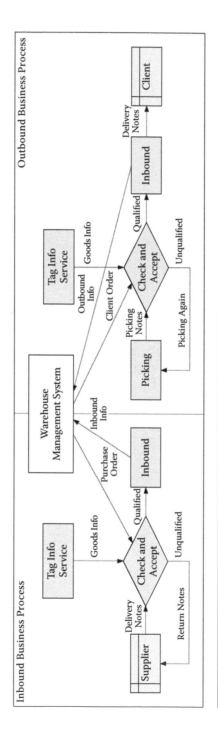

Figure 9.9 Inbound and outbound process.

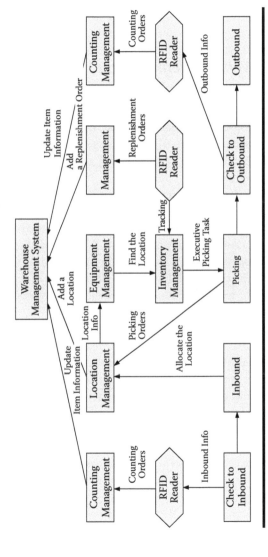

Figure 9.10 Intelligent goods location tracking management process.

In reference to the outbound process, the backend warehouse management system will arrange the outbound task and ensure that the goods are sent to the outbound area. In the outbound area, the goods are tracked by RFID readers and sent to the packing area to complete the packing task with the permission of the warehouse management system. By the means of monitors throughout the warehouse and the RFID readers fixed on the shelves, it is easy to find the exact location and maturity of goods to complete the automatic replenishment functions.

9.4.2 *Automatic Vehicle Identification*

An automatic vehicle identification system is more suitable for large-scale manufacturers, with thousands of vehicles entering and leaving each day. Although an original vehicle monitor system can register the vehicles, it cannot locate the vehicles or check the transported goods. RFID-based solutions can integrate vehicle management, logistics management, and tracking and security management for those large-scale manufacturers.

RFID-based vehicle electronic licenses can be used to take the initiative to efficiently manage vehicles. With the deployment of RFID readers at main gates, both the information of the trucks and their goods can be checked according to the logistical task in the backend information system. With the deployment of RFID readers at the main cross roads, even the location and speed of a vehicle can be obtained, which makes the cost much lower than deployment of the satellite global positioning system (GPS) technology on each vehicle. The entire system architecture is shown in Figure 9.11.

A RFID electronic license is delivered to each vehicle that is authorized to enter the enterprise. The license embeds unique and untearable RFID chips, so it cannot be replaced with a false license. Moving vehicles can also be interrogated and examined. Enterprise security departments can collect vehicle information at different gates or places, and check this information with the logistic information management system. The unmanned sentries at gates connect with alarm systems to intercept suspicious vehicles, if necessary. With the help of wireless terminal equipment, safeguards can also be set up at temporary checkpoints. Any vehicle passing the collection areas in an enterprise will be recorded. According to records in the information center, any suspicious vehicles can be checked and reported to the police.

The data and information of the RFID-based vehicle tracking system are collected automatically, without manual intervention, thus it can reduce the workload of safeguards. And, on the other hand, it can effectively avoid the possibility of collusion. As extension applications, tracking transport vehicles can also be used for tracking goods and getting the current location of possible stolen goods, which is extremely important in preventing the further spread of fake and shoddy products.

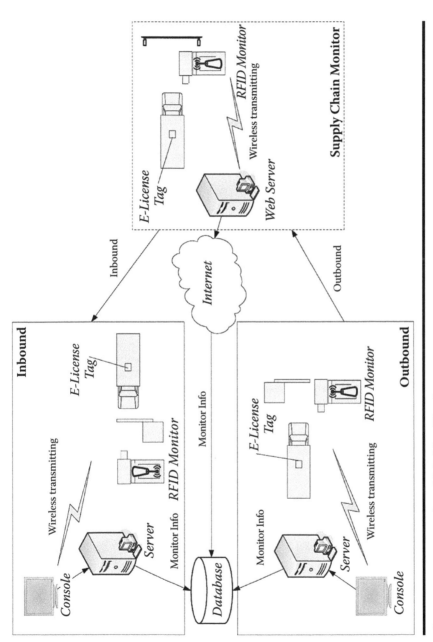

Figure 9.11 Automatic vehicle identification system.

9.5 Conclusion

The application of RFID systems to assets and vehicle tracking in the enterprise above is significant to logistics management. Compared with traditional barcode identification technology, RFID possesses characteristics, such as reading multiple tags once, reading/writing many times, and so on. In logistics management, information accuracy and timeliness are the key concern, which is an advantage of RFID-based applications. RFID technology enhances the degree of automation in goods sorting and reduces error ratio, causing the entire enterprise management to be open and effective. RFID technology reduces theft in the enterprise and gives reasonable reference to the scheduling of vehicles.

In the future we need to perfect the system to handle the challenges listed below:

1. It is necessary to prevent unauthorized insertion into the private information on the RFID tags. In the wireless channel between tags and readers and that of intelligent terminals and management system, there is a lack of a security solution to avoid remote eavesdroping on RFID tag information.
2. In the vehicle tracking system, it is credible to manage the vehicles with the RFID license tag. However, the vehicles with false license tags are invisible to the system since they don't have RFID data. As a result, it is important to make the license tag visible to the naked eye or develop more advanced scientific methods to solve this problem.

With the depth and extent of development of RFID technology, it is expected that RFID equipment will be multifunctional, small, with rapid transmission of information, etc. Meanwhile, the Internet of things, that support global RFID information exchange will enlarge the scope of RFID-based asset tracking.

References

[1] RFID Journal. http://www.rfidjournal.com/, accessed on 10/01/2007.
[2] EPCglobal. http://www.epcglobalinc.org/, accessed on 10/01/2007.
[3] Gary M. Gaukler, RFID in Supply Chain Management. Ph.D thesis, Standford University, 2005.
[4] Lai Fujun, Hutchinson Joe, and Zhang Guixian, Radio frequency identification (RFID) in China: opportunities and challenges, *International Journal of Retail and Distribution Management*, 33, 2005:905–916(12).
[5] Magnus Holmqvist and Gunnar Stefanssone, Mobile RFID—A case from volvo on innovation in SCM, *Proceedings of the 39th Annual Hawaii International Conference on System Sciences 2006* (HICSS '06), Kauai, Nawaii vol. 6, January 2006.
[6] Zhang Ting, Xiong Zhang, and Ouyang yuanxin, A framework of networked RFID system supporting location tracking, *2nd IEEE/IFIP International Conference in Central Asia on the Internet*, 2006, Tashkent, Uzbekistan, pp. 1–4.

[7] R. Angles. RFID technologies: supply-chain applications and implementation issues, *Information Systems Management*, 22 1:51–65, 2005.

[8] Nova Ahmed, Rajnish Kumar, Robert Steven French, and Umakishore Ramachandran, RFID: A reliable middleware framework for RFID deployment, *Proceeding of IEEE International Parallel & Distributed Processing Symposium* (IPDPS 2007), Long Beach, CA, March 2007.

[9] Young-Jun Seo, Deuk Kyoung Oum, and Peom Park, Bridging the real world with the RFID phone—Focus on development of innovative man machine interface and abundant wireless Internet service, *Proceedings of the 6th IEEE International Conference on Computer and Information Technology*, Seoul, Korea, September 2006.

[10] Yi Tianchen, Ai Minyu, Raymod Lai N, and Dou Xuechen, Application of WLAN in JTAV, *Proceeding of IEEE International Symposium on Microwave, Antenna, Propagation and EMC Technologies for Wireless Communications*, Beijing, China, August 2005.

[11] R. Kalden, I. Meirick, and M. Meyer, Wireless internet access based on GPRS, IEEE personal communications, April 2000.

[12] Raj Bridgelall, Enabling mobile commerce through pervasive communications with ubiquitous RF tags, *Proceeding of IEEE Wireless Comunication and Networking Conference 2003* (WCNC 2003) New Orleans, LA, March 2003, 3:2041–2046.

[13] B.S. Prabhu, Xiaoyong Su, Charlie Qiu, et al., WinRFID: A middleware for the enablement of radio frequency identification (RFID)-based applications, *Wireless and Sensor Networks: Technology, Applications and Future Directions*. John Wiley & Sons, 2005.

[14] Evan Welbourne, Magdalena Balazinska, et al., Challenges for pervasive RFID-based infrastructures, *Fifth Annual IEEE International Conference Pervasive Computing and Communications Workshops* (PerCom Workshop 2007), White Plains, NY, 388–394, March, 2007.

Chapter 10

RFID Enabled Logistics Services

Zongwei Luo, Edward C. Wong, C.J. Tan,
S.J. Zhou, William Cheung, and Jiming Liu

Contents

In this chapter, we introduce the concept of trustworthy pervasive services for enabling next generation E-logistics, the key enabler for improving and streamlining logistics operations. In pervasive services-enabled E-logistics, trustworthiness establishment is usually under resource constraints, presenting challenges for mutual trust establishment as well as data dependability. Thus, we will focus on security for mutual trust and data dependability aspects of pervasive service trustworthiness. We will present mutual authentication protocols for mutual trust establishment and a collaborative model for data dependability resolving data collision problems in pervasive service environments.

10.1 Introduction

As the world is becoming increasingly interconnected and interdependent, logistics rise up as the fundamental enabler for increasingly global economy connecting worldwide participates involving suppliers, factories, warehouses, distribution centers, retailers, and service and solution providers. As the Pearl River Delta (PRD) continues to grow as a manufacturing center for the world, Hong Kong has the opportunity to remain and grow as a logistics and enterprise headquarters hub. At the same time, Hong Kong is faced with mounting pressure to effectively manage and control disparate, international, and increasingly complex logistics networks and processes.

As a result of this increasing globalization of trade and manufacturing, system integration for logistics and supply chain management have garnered the interest of both researchers and practitioners. To enable global trading and manufacturing, there has been a renewed interest in devising and putting into practice integrated systems to consolidate all the isolated logistics systems. Thus, an integrated and developed environment is foreseen for enabling such a worldwide logistics and supply chain network development.

Logistics industry has been and continues to look for information technology (IT) systems and solutions to bridge three flows (physical goods, cash, and information), streamlining the supply chains by integrating transportation, distribution centers, retailers and manufacturers. Built upon the networked logistics systems, modern logistics demand more and more professional logistics services in order to improve profitability and competitiveness. These logistics services diversify and penetrate into all domains of logistics management in order to enhance resource utilization and reducing operational cost [1]. As evidence of this trend, IDC research reported that a third of logistics services segments are keeping involved and being refined [1]. We predict, in the future, that current networked logistics will no doubt be involved in the next generation of E-logistics dominated by pervasive logistics services.

This new pervasive logistics services paradigm promises more flexible and adaptable logistics; available anywhere, anytime, and in any means, these pervasive logistics services ride upon networked logistics systems, making it possible for organizations to better align IT resources with business priorities. This, in turn, demands seamless and/or dynamically created trustworthy communication and interconnections among logistics participants. In this chapter, we will illustrate this pervasive logistics service vision and present a trustworthy pervasive service framework for logistics.

To accomplish this, an architecture supporting the evolution of system development and easy integration of new and legacy systems is needed. Thus, we propose the adoption of Service-Oriented Architecture (SOA), utilizing technologies, such as Web service-based infrastructure, repository, and application development, to develop a trustworthy environment. This will enable the design, implementation, and control of logistics information essential to the delivery of logistics systems.

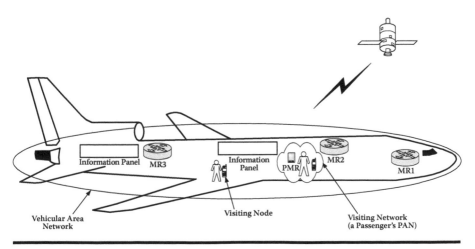

Figure 10.1   Interconnected networks for pervasive services.

The chapter sections are as follows:

■ 10.2: Pervasive Services Trend
■ 10.3: Open Issues and Opportunities
■ 10.4: Trustworthy SOA
■ 10.5: Case Study—RFID-Enabled Logistics Services
■ 10.6: Mutual Authentication for Mutual Trust
■ 10.7: Data Collision and Dependability
■ 10.8: Conclusion

10.2 Pervasive Services Trend

Pervasive services are emerging as the next paradigm for distributed and mobile computing, which are seamlessly available anywhere, anytime, and in any format. Logistics enterprises are also evolving towards the SOA trend, to offer logistics services over logistics networks. There is a need for mechanisms to connect the logistics networks and applications, and to enable logistics services over the logistics networks and allow access to services provided by these networks. The pervasive logistics services mechanisms will eventually allow dynamic and cost-effective creation of interconnections of disparate logistics network systems and processes resulting in improved efficiency in networks and applications development and deployment.

Logistics industry is pressing for such solutions, e.g., for realizing multivendor, multiplatform interoperability. Challenges are to reduce the time, risk, and cost of interoperability between disparate applications and devices from multiple vendors on multiple platforms. In light of these challenges, Fujitsu [2], for example, has been developing a SOA-based pervasive retailing framework for enabling retail store operations.

The pervasive logistics services vision has received strong support from the industry. Radio Frequency IDentification (RFID) technology, the pervasive infrastructure enabler for global supply chain visibility, is predicted to revolutionize the modern supply chain and logistics operations. With the introduction of Electronic Product Code (EPC) and as the global RFID standards mature, more and more companies are preparing to leverage RFID to integrate it into their supply chains for a competitive edge. RFID/EPC, the technology for the next generation of "Internet of Things," provides opportunities for new applications, services, and solutions over the next generation Internet. We see the trend of convergence of computing and communication, requiring for advanced service enabling technologies with mobility support, and service differentiation and secure communications.

Moving toward this pervasive vision, research and development activities are visible and yield substantial results in pervasive computing. Early work includes providing capabilities for dynamic networking of devices. IBM® has offered a framework to allow a diverse set of devices to connect via open network standards

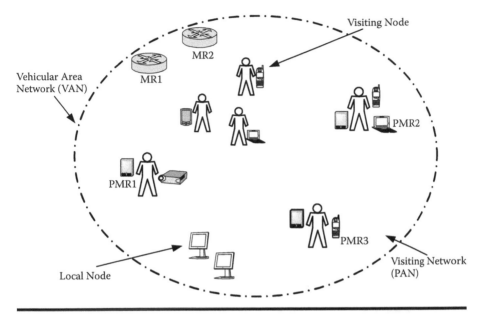

Figure 10.2   Pervasive service access.

to enterprise data and applications anytime and anywhere [3]. Microsoft® also planned to form a platform that promises to deliver convergence among disparate communications technologies [4]. Recently, Cisco and IBM planned to promote industry adoption of unified communications and collaboration, an important step toward enabling pervasive services [5].

10.3   Open Issues and Opportunities

In the pervasive services environment, it is predicted that users tend to be mobile and typically access resources and services using various tools and devices. Key challenges include the managing of security and of access control. As the logistics service becomes pervasive, trustworthiness of these services is becoming increasingly important as trust exists pervasively in the dynamic interactions and collaborations among logistics participants. Thus, in the following, we will identify research issues in developing trustworthy pervasive logistics services via reviewing related works.

10.3.1   Trustworthiness Research

Substantial research activities are reported in the domain of trust ranging from direct experiences and recommendations approaches to other dimensions, such

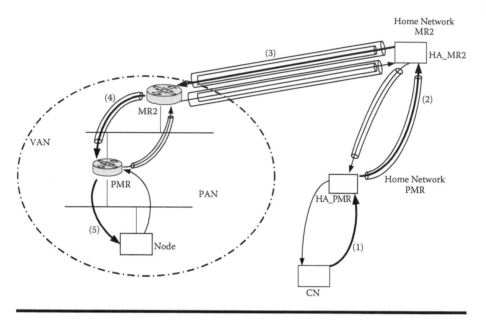

Figure 10.3 Secure channels for data access.

as security. Recommendations for managing context-dependent and subjective trust is described in [6]. Distributed trust based on direct monitoring is also proposed in combination with utility and importance of the situation [7]. In [8], the authors provide a rating process for trust. Distributed trust models for managing trust with quantitative a scheme using distributed recommendations was reported in [7]. In [9], the authors described a formal model for trust formation, evolution, and propagation. Game theory and distributed algorithmic mechanism design is reported in [10] to calculate trustworthiness based on reputation. Decision modules were also designed and integrated for assessment [11–13].

Research activities on security trust aspects are also reported. Frameworks resistant to attack for building up trust systems can be seen in [14–16]. Proposals for intrusion detection system (IDS) by extending it to an ad hoc wireless environment can be found in [17]. Distributed authentication mechanisms [18] are needed for devices, which cannot be trusted unless there is first-hand evidence.

Industry development on trust in pervasive services has been centered on Web services security and trust management. In 2002, VeriSign unveiled a trust services framework to enable trusted Web services [5,7] with support from major players including BEA®, HP®, IBM, Microsoft®, Oracle®, Sun Microsystems®, and webMethods®. Web Services Trust Language (WS-Trust) [19] provides extensions to WS-Security to secure a communication between two parties for trust relationships. It represents the industrywide efforts in Web Services trust. In WS-Trust, the two parties must exchange security credentials; however, each party still has to determine if they can trust the asserted credentials of the other party.

10.3.2 Issues and Opportunities

In summary, while many proposals exist for building trust pervasive services, there are various issues related to those proposals, such as:

- Many trust models, as well as frameworks, lack adaptability, the key to supporting pervasive services trustworthiness. As pointed out by many researchers, they were often designed for specific scenarios. This makes these proposals difficult to be deployed in different environments.
- They lack methods on how the trust models and frameworks can be applied in various scenarios. Solutions are needed to build trust relationships that can be used in evaluating trust for various applications.
- Many trust models and frameworks cover comprehensive elements, which are quite useful in trust evaluation. However, these elements are scattered in many different dimensions, lacking interconnections and interdependency analysis between them. There needs to be a demonstration on how a trustworthy pervasive service can be built out of the interdependent elements in the trust models.

Essentially, while we need a holistic view of integrating various trust elements and form a trustworthy environment, we have to bridge the gaps between the various trust models proposed, realization of various trust elements, and systematic deployment of trustworthy solutions for a trustworthy environment. We will address the issues by developing a service framework to allow transparent trustworthy pervasive services development. Further, to illustrate the framework feasibility, we will, using sensing networks services as an example, focus on the realization of security and dependability of aspects of pervasive service trustworthiness.

10.4 Trustworthy SOA

Pervasive logistics services are often based on SOA, an architecture supporting the evolution of system development and easy integration of new and legacy systems. SOA methodologies help enable more flexible and responsive IT environments to support the demands of pervasive logistics services.

In critical and ubiquitous infrastructure-enabled logistics and supply chain management-related processes and services, resource constraint has been a key concern in developing secure pervasive services. We will present a realization for a lightweight security protocol for mutual trust among pervasive services. To further enhance the mutual authentication protocols, a concept is presented for a collaborative model for data dependability resolving data collision problems in pervasive service environments.

10.4.1 Trustworthy Service Framework

We develop Trustworthy Service-Oriented Architecture (SOA-Trust) for the next generation logistics network infrastructure to streamline the pervasive logistics services development and delivery, technology deployment, localization, and adoption. A SOA-Trust framework includes three layers of abstractions: (1) the trustworthy service models, (2) the trustworthy interactions, and (3) service realization proxies. SOA-Trust extends the work in [20], including a metamiddleware service framework for autonomous service provisioning, which provides secure communication for mutual trust and data dependability.

SOA-Trust provides brokerage functions necessary in finding and using services provided by the middleware framework. To meet the challenges, such as heterogeneity and usability for pervasive service brokering, we use service description technologies and annotation languages, like Web Service Description Language (WSDL), to provide an information abstraction layer for the pervasive services brokerage. It describes and correlates pervasive services provided through the WSDL and annotation, faciliating assurance of the functional consistency of the pervasive services provided by different system developers.

The SOA-Trust brokers make the decisions involving service resources over multiple pervasive services. The core module is the brokering engine driving the orchestration engine to execute the brokering plans with brokering logics (algorithms) and context. Depending on the different nature of the middleware service request, the service broker in turn invokes one of the specific brokers to help fulfill the service request.

The SOA-Trust will provide services for secure event processing to support streaming capabilities. It abstracts the event processing functions into messaging, logics, and context. These functions are concerned with more than data reduction operations, such as filtering, aggregation, and counting. Functions, such as mutual authentications and data dependability, are provided for mutual trust among pervasive services. Each of these functions will be the features supported by the middleware as well as functional components published in the SOA-Trust service framework.

The processing messaging mainly concerns the message streaming protocol. The process logics contain algorithms to handle the message streams for filtering, aggregation, counting, etc. The process context establishes the trustworthy relationships for the streaming patterns among the event processing modules.

10.4.2 Design Rationale

SOA-Trust intends to provide flexibilities to open up opportunities for effectively reconfiguring and orchestrating pervasive services at a metalevel for supporting more agile and scalable development of distributed systems. The SOA-Trust framework that is building up the pervasive logistics service infrastructure is characterized as:

■ SOA-Trust is built upon a foundation of global standards, best practices, and methodologies of service-oriented computing. Leveraging SOA, it provides evolutionary architecture for building the pervasive logistics services infrastructure. SOA-Trust offers a provision framework for new components, technologies, and services, and facilitates the solution and system deployment.

■ SOA-Trust ensures the consistency in finding and using information and services in the logistics networks. In SOA-Trust, resources are made available to other participants in the network as independent services that are accessed in a standardized way. This provides for more flexible loose coupling of pervasive service resources.

SOA-Trust intends to help build logistics services to address the following challenges:

■ Challenges in adopting present and upcoming standards (e.g., RFID, EPC), which are still evolving.
■ Not prepared for adaptation with new components, technologies, and services, and is limited in its ability to reuse existing components and services.
■ Challenges in its consistency to find and use information and services.
■ Challenges in legacy application and data integration to preserve the organizations' IT investment when they migrate to pervasive infrastructure.
■ Not scalable and not prepared for large-scale pervasive technology deployments.

SOA-Trust offers service framework to turn an organization's existing and new applications and data sources into a pervasive logistics services infrastructure. It aids technology in localizing and adapting to create a more modular and flexible set of interfaces for the ongoing conversion to pervasive services. SOA-Trust enables evolution of pervasive solutions over time. It provides important benefits for scalable pervasive service implementations and system deployment. SOA-Trust enables end-to-end solutions that can help achieve improved business efficiencies by applying pervasive services to logistics business processes in the related business logics.

10.4.3 Pervasive Logistics Services Enabling Capabilities

SOA-Trust promises an effective and trustworthy approach to seamless integration and orchestration of distributed resources for dynamic business processes along the supply chains. The integration and adaptation needs of next generation E-logistics are addressed by a middleware integration framework [20,21] in SOA-Trust. This service-oriented intelligent middleware service framework fulfills the needs of how to embed the autonomy-oriented computing paradigm in the framework to enable autonomous service brokering and composition for highly dynamic integration among heterogeneous middleware systems [20,21].

10.5 Case Study: RFID-Enabled Logistics Services

Recently renewed interests in RFID has turned logistics industry's attention to potential benefits from real-time logistics information visibility. The up-to-the-second or -minute information status of available inventory and goods delivery supports better logistics services and planning. Also, delivery vehicles installed with global positioning system (GPS) sensors allow their locations to be accurately tracked for more effective fleet management. While the involved sensor technologies have been becoming more mature in recent years, integrating them together for building E-logistics infrastructures and networks demands middleware frameworks that enable different parties to communicate and collaborate in an efficient and effective manner. SOA-Trust helps to provide the following functions in order to take full advantage of what RFID can promise:

- **Reactivity:** RFID and EPC technologies have promised real-time global information visibility for E-logistics participants. To benefit from such visibility, the E-logistics participants will be able to identify the interested situations and react to each situation when it happens. The event will be reported and notification will be sent to interested E-logistics participants.
- **Integration:** System integration and maintenance burdens increase with the number of choices of the RFID middleware systems available to interface with the E-logistics systems. The challenges posed by such middleware proliferation and heterogeneity as well as how dynamic logistics environment impose certain architectural and functional requirements are addressed in SOA-Trust [20].
- **Adaptability:** SOA-Trust enables evolution of pervasive solutions over time. It adopts the autonomic service provisioning capabilities to further enhance its adaptability for scalable pervasive service implementations and system deployment [20,21].

However, there are still other challenges that need to be addressed in the process to fully develop trustworthy RFID-enabled logistics systems and solutions.

10.5.1 Resource Constraint

The low cost demanded for RFID tags, the power restrictions, and being nonresistant to physical tampering are common limits in the RFID system. Therefore, most traditional security schemes widely used in Internet or existing applications are not feasible in RFID environments. Some lightweight and resource-saving security mechanisms must be imposed on the RFID tags and readers to address the authentication and privacy problems.

10.5.2 Implementation Hardship

The low cost demanded for RFID tags causes them to be very resource-limited. The power restrictions and the nonresistance to physical tampering also make RFID systems vulnerable to passive and active attacks. Moreover, tags being short of cryptographic functions makes the traditional security schemes that is widely used in Internet or existing applications unfeasible in RFID environments as well. According to [20,23], typical tags that can only store hundreds of bits of data, roughly have between 5,000 and 10,000 logic gates. Within this gate counting, only between 250 and 3,000 gates can be devoted to security functions, including random number generation, encryption or decryption, hash operation, etc. This limited resource results in popular cryptographic algorithms, including Data Encryption Standard (DES), Advanced Encryption Standard (AES), public cryptographic algorithms (e.g., RSA and ECC), and cannot be easily implemented in low-cost tags. For example, it was reported that for a standard implementation of the AES between 20,000 and 30,000 gates are needed [20,23].

10.5.3 Privacy Concerns

Powerful track and trace capabilities promised by RFID tags create new threats to user privacy. The bogus identity also increases risks for RFID applications. As a result, some light weight and resource-saving security mechanisms must be imposed on the RFID tags to address the authentication and privacy problems in RFID and in [22,23], authors proposed a few privacy solutions. In [24,25], the authors provided detailed discussions on a XOR-based, lightweight mutual authentication RFID protocol. The primary advantage of the protocol proposed in [24,25] is that it only needs simple Hash operation, XOR, and random number generation for both tag and reader (or backend system). This simple function can be implemented easily in passive tags with little incremental cost. Another strong point of the protocol is that the read/write access is integrated into the protocol. This makes it very suitable to the applications in which each tag has many restricted access memory blocks. Its resource-saving feature, scalability, and high performance meet the requirements in metering and payment of E-commerce applications.

10.5.4 Data Collision

Many readers in a RFID application form a reader network through which readers can send the data to a backend system. However, reader collision problems cannot be avoided in such multireader RFID environments. Unlike the tag collision problem, which has been widely discussed in existing research, the impact of reader collision is not regarded as important as tag collision because it seldom required multireaders in the past few years. For improving read rate and correctness, several readers are put together to form a dense RFID reader environment. In a dense reader environment, reader collision is also becoming a key issue for the application of RFID. The current RFID reader network is a centralized model in which any message is streamed to a single, central backend system.

10.5.5 Our Demonstration

In the following sections, realizations for the trust elements, such as mutual authentication for mutual trust and data dependability, are provided to support trustworthy RFID-enabled logistics system and application development. These trust elements will be orchestrated and invoked by the brokerage systems in the SOA-Trust to form end-to-end supply chain solutions.

The scenario of the demonstration is as follows:

- In RFID-enabled pervasive logistics services, RFID readers will communicate with tags and other readers.
- In this RFID network, trust relationships have to exist between the RFID readers and tags.
- Before the communication begins, pervasive logistics services participants (i.e., readers and tags) have to authenticate with each other to establish corresponding trusted relationships.
- However, there are risks in the trusted relationship established only through the very first "say hello" authentication. For example, a Denial of Service (DoS) attack resulting from data collision would pose a threat to data dependability in the communication. Thus, trust mechanisms are needed for data reliability to maintain the trusted relationship.
- The authentication protocol and data dependability solution are trust services developed in the SOA-Trust framework.

10.6 Mutual Authentication for Mutual Trust

Although there are many authentication protocols already proposed, their implementation is not satisfactory when deploying on a real RFID system. The simple authentication protocol in RFID is based on a tag's physical feature. Tags themselves have no access control function, thus, any reader can freely obtain information from them. As a result, an authentication (as well as authorization) scheme must be established between the reader and the tag so as to achieve the privacy issue of a RFID system. Since the communication between a tag and a reader is by radio, anyone can access the tag and obtain its output, i.e., attackers can eavesdrop on the communication channel between tags and readers, which is a cause of consumer apprehension. So, the authentication scheme employed in RFID must be able to protect the data passing between the tag and the reader, i.e., the scheme itself should have some kind of encryption capability.

RFID readers may only read tags from within the short tag operating range (e.g., 3m), the reader-to-tag (or forward channel) is assumed to be broadcast with a signal strong enough to monitor from long range, perhaps 100. The tag-to-reader (or backward channel) is relatively weaker and may only be monitored by eavesdroppers within the tag's shorter operating range. Generally, it will be assumed that eavesdroppers may only monitor the forward channel without detection.

10.6.1 Lightweight Mutual Authentication Protocol[24]

Before presenting the overall protocol, we make the following assumptions. First, it is assumed that each tag has k readable and writable blocks. Second, each tag stores the following information:

- Tag identification is used to identify a tag.
- Transaction counter c with initial value of 1 will be increased by 1 after each successful transaction.
- R is the randomly generated number.
- W is the workable area.
- The secret is shared with the backend database or system with the initial value.

We also assume that each tag shares two public hash functions, H and G, and with the backend system.

In the following, the lightweight challenge-response protocol [24,25] adapted to the limited resource criteria of a passive tag is illustrated as below.

10.6.1.1 Initialization

Assume a passive tag contains three workable regions (e.g., W1, W2, and W3) opened for both reading/writing. At the very beginning, beside the EPC code, each tag comes with an initial secret information, s0, and a transaction counter, c (usually a 8 to 20 byte number depending on the output length of the hash function G), which will be used by the backend for tag verification. On the other hand, the backend stores lists of information on each tag that it can work with. That piece of information basically contains a tag's initial secret information s0 and workable region(s) Wk, where k belongs to 1, the number of workable regions.

10.6.1.2 Protocol Procedures

- For Ith transaction, the backend sends a read/write request to the tag.
- The tag responds with both the transaction counter value c_i and the hashed output $G\{s_i\ c_i\}$ back to the backend.
- From the received transaction counter c_i, the backend first does hashing s_0 for c_i times in order to get sb_i and then calculates the hash output $G\{sb_i$ XOR $c_i\}$ by iterating each tag stored. Then, it compares the value between $G\{sb_i$ XOR $c_i\}$ and $G\{s_i$ XOR $c_i\}$ and the equality holds if and only if $sb_i = s_i$, which means that the remote tag is authenticated by the backend.
- Once finding the matching tag, the backend generates a random session number, R, and makes the output containing three elements, R XOR $G\{sb_i\}$, $G\{sb_i$ XOR R XOR $W_k\}$ and W_k, and sends them back to the tag.
- The tag can get back R_t from element R XOR $G\{sb_i\}$ by doing XOR with its internal hashed s_i. Then, it calculates the hash output $G\{s_i$ XOR R_t XOR $W_k\}$ to see whether the value is equal to $G\{sb_i$ XOR R XOR $W_k\}$ or not. Again, the equality holds if and only if $s_i = sb_i$ and $R_t = R$, which means that the backend is also authenticated by the tag. At this point, the mutual authentication procedure is completed and a common random session number, R, is being agreed and setup between the tag and the backend for data encryption. It should be noticed that the authorization is also achieved since W_k defines what regions are opened for reading/writing within this transaction.
- Since mutual authentication is established, it is ready to begin data transfer between the tag and the backend. To protect the data passing inbetween, the random session number, R, is employed as a key for doing the encryption for that purpose. For each data to be transferred, the sending party (the tag for read query case and the backend for write query case) produces two elements, $G\{R\}$ XOR data and $G\{data\}$. On the other hand, once receiving these elements, the received party (the backend for read query case and the tag for write query case) can retrieve back the data by doing XOR with $G\{R\}$ XOR data and then verify its value by hashing it and comparing it with $G\{data\}$.
- Finally, the tag will increment the transaction counter, c_i, by 1 and update the secret information from s_i to $s_i + 1$ by hashing it using H function after current data transfer phase is completed.

10.6.2 Related Work

A unique serial number is a popular means for developing authentication protocols. In this scheme, each tag must hold a PIN or PIN list. The PIN-based authentication is similar to kill and sleep [26] in which each tag stores a unique PIN in its local memory. While authenticating with the reader, the tag will send its PIN to the reader for authentication. Unlike the PIN-based scheme, in the PIN list-based authentication, which is also similar to minimalist cryptograph method addressed in [27], every tag stores a list of PINs and each time it will randomly select a PIN from the list to authenticate with the reader. However, both the replay attack and the physical tampering can easily breach the authentication. Moreover, no mutual authentication is provided in unique serial number-based authentication protocol.

Hash-based authentication protocol is similar with our scheme. Different approaches include hash-lock [28,29], semirandomized hash-lock [30], randomized hash-lock [31], and hash-chain [32]. Other hash-based solutions are also discussed in [33–36]. All of these protocols rely on synchronized secrets residing on the tag and backend server and require a one-way hash function from the tag. These approaches show how guaranteeing the untraceability by updating tag identifier would increase the workload of backend servers. Compared with our hash-based protocol, these existing authentication protocols do not integrate reading and writing requirements with authentication.

The challenge–response-based authentication protocols are also similar with our scheme. Many different transformation techniques were used to provide security during the challenge or response process. The transformations include asymmetric cryptography, XOR [22,37], hash function [38–40], and physical unclonable functions [41]. Like existing hash-based schemes, current challenge–response-based authentication protocols do not provide the integration properties of reading and writing with authentication.

Some other authentication protocol in RFID include HB (Hopper and Blum) and HB-enhanced [42,43], a list of licit readers approach, noisy approach [44], silent tree walking and backward channel key negotiation [28,29], and active information exchanging [45]. However, these existing solutions are only feasible in some given RFID applications and not general to any RFID application. The security of these protocols is not yet proved theoretically. For the detail of these existing protocols, readers can refer to [46].

10.6.3 Further Discussion

To summarize, most of the above methods are not likely to be a fully satisfactory solution. Thus, it seems imperative to explore alternatives to get a workable scheme. While the lightweight mutual authentication proposed in [24,25] has the strength of a resource-saving feature, scalability, and high performance, it is possible that attackers can capture and log the authentication messages by sniffing the communication between tags and readers. Then, attackers, in personating the identity of the legal tag, can replay the message of the authentication protocol to the backend system. Although this type of attack could be easily mitigated due to the randomly generated information to protect the exchanged messages, it could lead to potential DoS attacks. That is, attackers can sniff the communication and resend the message to the party. In this way, it is possible to put the system out of service. Attackers can also destroy the synchronization of the security protocol, pose which can a DoS risk to the RFID systems.

DoS attacks seem hard to mitigate. Only if there are strict access controls in which sniffing, message replay, or jamming is not possible, a DoS attack risk can be minimized. A mechanism for checking the message received and message filtering, i.e., data filtering and anticollision, are also methods to reduce this risk. Thus, in the following section, we will discuss techniques to avoid data collision to help mitigate the DoS attack.

10.7 Data Collision and Dependability

In this section, we will focus on reader collision problems by enhancing the data dependability. It relies on a collaborative peer-to-peer network for solving reader collision problems in dense RFID reader environments [47]. It is a fully distributed and self-organized overlay network. The model can be used in a dense reader environment to exchange information directly among readers without any central control or intermediation. In RFID systems, the interference refers to the collision of different radio signals with the same frequency, which will lead to distorted signals. Occasionally, if tags or a reader broadcast the same frequency radio signal, collisions will happen. The RFID internal interference, (RFID collision problem) contains tag-to-tag, tag-reader, and reader-to-reader collision.

10.7.1 Dense RFID Reader Environment

Dense RFID reader environment refers to the RFID application scenes in which multiple readers are deployed in a predetermined place to provide highly reliable and correct reading of tags. In regular RFID environments, one reader is enough to scan and read multiple tags. However, a single reader sometimes cannot provide high read rate while keeping the correctness of reading if multiple tags stream into the reader's reading zone; for example, misreadings and crashed reading results are common in such a RFID application. For the purpose of improving the read rate and correctness, several readers are put together to form a denser reader environment.

10.7.2 Data Collision Problem

Reader collision problems are rooted in the limited frequencies available to RFID readers and tags. Unlike other wireless communication techniques, although RFID operates at different frequencies and the choice of frequency depends on the application, it is not a free choice as radio frequencies are strictly regulated. Regulations are, of course, needed in order to avoid interference between the different radio systems. In order to provide interoperability worldwide, a variety of specific frequencies for RFID have been declared. These are known as ISM frequencies (Industrial–Scientific–Medical) [48]. The reason for this is that they were gratis in most countries and represented a selection of low, intermediate, and high frequencies, allowing for RFID systems with different purposes. Therefore, the limited available frequencies prevent separating each reader to operate under different frequencies. At the same time, low-cost tags are also short of selective frequencies, which also leads to reader collision.

Other than the above factors, we also believe the lack of information exchanges among RFID readers inevitably sharpens the circumstance of reader collision. While there is no direct information exchange among readers, the reader's current status is not fully utilized by other readers. First, readers can collaborate with each other to reduce the misreading and enhance their reading performance. Second, readers can work together to reduce or eliminate reader collision in dense reader environments. Finally, readers can exchange the reading list, a list of tags being read, with other readers to reduce redundant reading. All together, if an infrastructure is available for readers to exchange information directly, it is likely to reduce or eliminate reader collision.

10.7.3 Proposed Anticollision Schemes

To the best of our knowledge, the pioneering work about multireaders and their related reader collision problem can be found in [49,50], and a short survey of RFID collision, especially tag collision, was given in [51]. In [49,50], the authors for the first time addressed the RFID interference and reader collision problem. In theory, the reader collision problem is equivalent to the co-channel frequency assignment problem, which, in turn, is equivalent to the simple graph coloring problem. Because the graph coloring problem is a well-known NP problem, the author addressed two approximate methods: centralized approximate algorithms and local approximate algorithms. In [52], the authors addressed two enhanced graph coloring based schemes to schedule the various RFID readers in the network.

The anticollision methods mentioned in [49,50] and their enhancing versions are useful in small scale and static RFID environments. One disadvantage of the centralized approximate methods is the need of predistribution of the allocation. Hence, they cannot be generally used in RFID applications. The distributed greedy algorithm is the one which is most similar to our dynamic anticollision scheme. However, they do not address how to exchange the frequency each reader uses. To the contrary, we not only build an interconnected reader network to distribute the reader's information, but also construct a dynamic anticollision scheme. In general, both the high computation and the inadaptability limit coloring approaches in RFID applications.

Waldrop et al. [53,54] are also the first works (Colorwave) to address reader collision problem. In particularly, it is a time division multiple access-based (TDMA) anti collision algorithm. In Colorwave, first a reader collision network must be constructed. Then the reader collision problem is reduced to a classic coloring problem in graph theory, which is the same used in [49,50]. PULSE [55,56] is another TDMA-based anticollision scheme. In PULSE protocol, it assumes that there are two separate communication channels: data and control.

Listen Before Talk (LBT) is a Carrier Sense Multiple Access-based (CSMA) anticollision protocol defined in ETSI EN 302 208 [57], which is an evolving standard being developed for RFID readers. Like normal CSMA, in LBT, before querying the tag, the reader first listens on the data channel for any ongoing communication for a specified minimum time. The theoretical analysis for reader collision and LBT method appears in [58]. The CSMA-based anticollision approach is addressed in [59].

The Class I Generation 2 UHF standard [53] addressed by EPCGlobal [60,61] uses spectral schemes, i.e., it is based on FDMA (Frequency Division Multiple Access). It separates the reader transmissions and the tag transmissions spectrally such that tags collide with tags, but not with readers and readers collide with readers but not with tags. However, most of the low-cost tags do not have frequency selectivity. Hence, when two readers using separate frequency communicate with the tag simultaneously, this leads to collision at the tags. Thus, multiple readers-to-tag interferences still exist in this standard.

Q-Learning [62] is a hierarchical, online learning algorithm that finds dynamic solutions to the reader collision problem by learning the collision patterns of the readers and by effectively assigning frequencies over time to the readers. This protocol maintains a hierarchical structure, which will require extra management overhead. Q-Learning assumes collision detection for readers. However, in our dynamic antischeme, we addressed how to detect collision in dense reader environments. The complicated structure also makes it unsuitable for general RFID applications.

Naive sending mentioned in [59] is similar to the random backoff addressed in [49,50]. In naive sending, each reader transmits a reader-tag inventory request if only if it needs to send. If either reader-to-reader collisions or reader-to-tag collisions occur, the conflicted readers must resend the query again at freely or at constant intervals. Naive sending is quite prone to both reader–tag collision and reader–reader collisions. Ransom sending [59] is an improved version of naive sending. In this method, randomization in sending and resending is used to reduce the chance of collision. Thus, if the readers choose to back off for a random interval before sending a read command, the probability of collision may be lower. Since the different readers choose to send data after random intervals, the sending from two readers would probably not be concurrent, given that they independently choose their resending interval.

10.7.4   Establishing the Collision Model

In order to eliminate or reduce reader collision, current reader networks can be modeled as a collaborative network for a reader to directly exchange information with other readers. Other than the reader collision, some other problems occurring in current dense reader environments can also easily be resolved if such an information exchanging infrastructure is available. Peer-to-peer computing (P2P) technology is a preferred distributed scheme for the collaborative network. P2P is also regarded as a distributed model for information exchange. Therefore, in this section we present a collaborative peer-to-peer network for the reader collision problem in dense RFID reader environments. This novel model is constructed on the RFID reader collision model such that it is a fully distributed and self-organized overlay network. Our new model can be used in dense reader environments to exchange information directly between readers without any central control or intermediation.

Although a simple reader collision network cannot resolve the problem of reader collision, we can make use of it to build a high efficient anticollision network. If two readers potentially interfere with each other, either reader-to-reader collision or reader-to-tag collision, we connect these two readers by an undirected edge. After building the reader collision network, a central server can periodically remove or disable some readers to avoid a reader collision problem, which can easily use the methods in [49,50]. However, a simple static schedule will result in a misreading problem in multitag and dense reader RFID applications. Moreover, reader's individual anticollision capacity cannot efficiently resolve the problem of reader collision in dense reader environments. If we can build an interconnected collaborative network, it would be beneficial to also design a high efficient anticollision scheme.

After the construction of a reader collision network (RCN), the interconnected RFID reader collision model (IRCM) can be built on RCN [47]. For the sake of stationary RFID reader networks, each reader can periodically update its collision neighbor and routing neighbor by the above algorithm. Therefore, it is unnecessary for additional update and maintenance protocol for IRCM. For a dynamic RFID reader network in which some readers may intentionally or unintentionally crash down, a separate update and maintenance protocol is helpful. We will take that it into account farther in our future research works.

Because of collision, including tag collision and reader collision, not all readers' queries will receive an anticipated response. For example, if a reader–tag collision occurs during a readers' reading period, no response will be sent to the readers because the interrogated tag does not know which reader is interrogating it. Similarly, during read-reader collision, tags are unaware of how to reply.

Some research programs have worked out a number of solutions for tag collision. Therefore, the tag collision can either be avoided or its impact on correct read could be precomputed before the deployment of a RFID system. In both cases, the impact cannot be changed. Although the impact of other environmental factors varies with different RFID applications, its effect will become more and more insignificant with the improvement of the quality of manufacture and the rationality of deployment.

10.7.5 Modeling the Collision States

The correct read rate of a reader during a given period is defined as Query-Hit Ratio, denoted as QT. Reader's Normal Query-Hit Ratio (NQT), denoted as QTn, is defined as the proximate value of QT while the reader operates without reader collision. By our definition, the value of NQT is always equal or less than 1(QTn = 1). In practice, ordinary NQT is always less than 1 because we cannot get rid of all the artificial impacts. The actual value of QTn varies with diverse readers. We also suppose that QTn for a given reader can be computed in experimental RFID environments. Generally, in practical RFID environments, we can safely assume that no reader occurs while QT belongs to the range of (QTn, 1). We also assume that if reader collision happens, the QT drops down as well. The lower the value of QT the more collisions that the reader encounters. Therefore, each reader can compare its current QT value with the normal QT value (QTn) to estimate if it will come up against a collision problem or not. If QT is less or equal to QTn, it speculates that some collisions will occur. By carefully computing the difference value between QT and NQT, a reader can also be capable of determining the deterioration of collision.

We also determined the extent of collision and semicollision. Reader's semicollision threshold (QTt) is defined as the lowest value of QT while the reader operates with semireader collision. For semicollision, we define a semicollision threshold as QTt. While the QT follows into the range of (QTn, QTt], the reader is in a semicollision state. If collision happens so frequently that misreading and other performances are greatly reduced, we can conjecture that reader collision occurs with high probability. Therefore, the reader changes to a disabled state, in which it stops interrogating tags and frees its whole reading cycle or frequencies for other readers. The threshold of disabled state is QTf. If QT is less or equal to QTf, a reader changes to a disabled state automatically. Therefore, a reader's disabled threshold (QTf) is defined as the lowest value of QT while the reader stops interrogating any tag. To sum up, each reader has the following states:

- Initialization state: Before the running of a RFID system, all readers remain in this special situation. Only when the RFID system starts running, will the reader change its state to a normal state.
- Normal state: When switched on, readers will go into the normal state in which they will query and read tags as if there is no collision. As long as QT is greater than QTn, a reader keeps itself in a normal state.

- Seminormal state: State in which some collisions do occur, but it is not so serious that it will result in numerous misreading. When QT drops below QTn but is not larger than QTt, a reader changes its state from normal to semicollision. While the collisionis mitigated and QT is larger than QTn, the reader will change back to a normal state.
- Collision State: State in which a serious collision has happened and results in slight misreading errors. Only when QT drops below QTt will the reader change its state from semicollision to collision. However, depending on the variety of QT, a reader will change its state to a disabled or semicollision state. If QT continuously drops below QTf, a reader stops its reading and changes to a disabled state. Contrarily, if QT goes up greater than QTt, the reader restores back to its semicollision state.
- Disable state: State in which the reader stops querying and reading tags. A reader keeps itself in this state for some time. If timeout occurs, it automatically resumes to collision state.
- End state (or stop state): If we power off the reader, it comes into a stop state and all tasks are ended as well.

10.7.6 The Anticollision Solution

In the process of developing dynamic anticollision schemes, we assume that each reader has a controllable reading speed, denoted as V, the number of reads per second. The reading speed decides how fast a reader can query and read tags within its vicinity. The difference between reading speed and query hit rate is that the latter refers to how successful the reading is. Actually, reader speed is tightly related to reader collision. In dense reader environments, high reading speed may result in highreader collision. For example, when two readers encounter a collision problem, if one reader reduces its reading speed, the probability of a collision will also decrease. Further, if one reader stops reading, the collision will disappear.

Accordingly, we can shift the reader into different states to force it to read with different speeds. This way, the collision can be reduced or avoided. Apparently, if any reader independently changes its state according to the above transition diagram, it has the same effect as naive send or random sending anticollision schemes [59].

Clearly three controllable and adjustable parameters (QTn, QTt, and QTf) could be used to control the transition. By adjusting these values, a reader will change to a different state. For improving the performance of anticollision, it is better for each reader to decide how to change its state by combining local and global states. For that reason, in our dynamic anticollision scheme, each reader periodically announces its local state to other readers though the interconnected RFID reader collision model (IRCD) constructed above. After receiving other readers' information states, a reader can adjust the value of the three controllable parameters synthetically to change its state and, thus, change its reading speed to avoid collision or reduce the probability of a collision.

10.8 Conclusion

In this chapter, a trustworthy pervasive service framework for E-logistics infrastructure and network integration and development has been introduced. It is designed to enable more responsive supply chains and better planning and management of complex inter-related systems. The proposed adoption of SOA can support the evolution of logistics information system development and easy integration of new and legacy systems.

We take a holistic view of integrating various trust elements to develop the SOA-Trust framework to bridge the gaps among the various trust model proposed, realization of various trust elements, and systematic deployment for a trustworthy environment. SOA-Trust supports the evolution of system development and easy integration of new and legacy systems. SOA methodologies help enable more flexible and responsive IT environments to support the demands of pervasive logistics services.

In a critical and ubiquitous infrastructure enabled logistics and supply chain management-related processes and services, resource constraint has been a key concern in developing secure pervasive services. We realize that there is a need for a lightweight security protocol for mutual trust among pervasive services. To further enhance the mutual authentication protocols, SOA-Trust addresses the data collision and dependability problem, yet another major concern. Also presented in this chapter is the desire for a collaborative model for data dependability resolving data collision problems in pervasive service environments.

References

[1] http://www.idc.com/getdoc.jsp?containerId=CN383107N
[2] http://www.fujitsu.com/us/news/pr/ftxs_20060116-1.html
[3] http://www-128.ibm.com/developerworks/web/library/wa-pvc/
[4] http://www.engadget.com/2006/06/26/microsoft-unveils-unified-communications-platform/
[5] http://www.eetasia.com/ART_8800455871_499495_2ba350a0200703.HTM
[6] A. Abdul-Rahman and S. Hailes. Using recommendations for managing trust in distributed systems. In *Proceedings of IEEE Malaysia International Conference on Communication, Kuala Lumpur, Malaysia*, November 1997.
[7] A. Pirzada and C. McDonald, Establishing trust in pure ad-hoc networks. In *Proceedings of the 27th Conference on Australasian Computer Science*, vol. 26, pp. 47–54, Dunedin, New Zealand, 2004.
[8] S. Buchegger and J.-Y.L. Boudec. A robust reputation system for p2p and mobile ad-hoc networks. In *Proceedings of the 2nd Workshop on the Economics of Peer-to-Peer Systems*, Cambridge, MA, June 2004.
[9] M. Carbone, M. Nielsen, and V. Sassone. A formal model for trust in dynamic networks. In *Proceedings of the 1st IEEE International Conference on Software Engineering and Formal Methods*, pp. 54–63, Brisbane, Australia, September 2003.

[10] H. Sun and J. Song, Strategy-proof trust management in wireless ad hoc network. In *Proceedings of the IEEE Canadian Conference on Computer and Electrical Engineering*, Niagara Falls, ON, 2004.

[11] N. Dimmock. How much is "enough?" Risk in trust-based access control. In *Proceedings of the 12th IEEE International Workshop on Enabling Technologies*, p. 281, Washington, D.C., June 2003.

[12] D. Quercia, and S. Hailes. MATE: Mobility and adaptation with trust and expected-utility. *International Journal of Internet Technology and Secured Transactions*, vol. 1, pp. 43–53, 2007.

[13] D. Quercia and S. Hailes. Risk aware decision framework for trusted mobile interactions. In *Proceedings of the 1st IEEE/CreateNet International Workshop on the Value of Security through Collaboration*, Athens, Greece, September 2005.

[14] J.R. Douceur. The Sybil attack. In *Proceedings of the 1st International Workshop on Peer-to-Peer Systems*, pp. 251–260, Cambridge, MA, March 2002. Springer-Verlag, Heildelberg, Germany.

[15] J. Liu and V. Issarny. Enhanced reputation mechanism for mobile ad hoc networks. In *Proceedings of the 2nd International Conference on Trust Management*, vol. 2995, pp. 4862, Oxford, U.K., March. 2004. LNCS.

[16] J.-M. Seigneur and C.D. Jensen. Trading privacy for trust. In *Proceedings of the 2nd International Conference on Trust Management*, vol. 2995, pp. 93–107, 2004. Springer-Verlag, Heidelberg, Germany.

[17] P. Kannadiga, M. Zulkernine, and S. Ahamed, Towards an intrusion detection system for pervasive computing environments. *In Proceedings of the International Conference on Information Technology (ITCC)*, pp. 277–282, IEEE CS Press, Las Vegas, NV, April 2005.

[18] H. Luo, P. Zerfos, J. Kong, S. Lu, and L. Zhang, Selfsecuring ad hoc wireless networks. In *Proceedings of the Seventh IEEE International Symposium on Computers and Communications* (ISCC'02), Taormina Giardini Naxos, Italy, pp. 567, 2002.

[19] http://specs.xmlsoap.org/ws/2005/02/trust/WS-Trust.pdf

[20] Z. Luo, J. Li, C.J. Tan, F. Tong, A. Kwok, E. Wong, and H. Wang. Intelligent service middleware framework. In *Proceedings of the 2006 IEEE Conference on Service Operation, Logistics and Informatics*, June 21–23, pp. 1113–1118. Los Alamitos: IEEE Computer Society Press.

[21] Z. Luo, M. Wang, W.K. Cheung, J. Liu, F. Tong, and C.J. Tan, An agent-mediated middleware service framework for e-logistics. Book Chapter in Advance Series in E-business (ASEB), 2007 (in print).

[22] P. Peris-Lopez, J.C. Hernandez-Castro, J. Estevez-Tapiador, and A. Ribagorda. RFID systems: A survey on security threats and proposed solutions. *International Conference on Personal Wireless Communications (PWC'06)*, Albacete, Spain, September 2006.

[23] P. Peris-Lopez, J.C. Hernandez-Castro, J.M. Estevez-Tapiador, and A. Ribagorda. LMAP: A real lightweight mutual authentication protocol for low-cost RFID tags, Workshop on RFID Security, Graz, Austria, 2006.

[24] Z. Luo, T. Chan, and J. Li, A lightweight mutual authentication protocol for RFID networks. *Proceedings of ICEBE 2005*, Beijing, China, 2005.

[25] Z. Luo, T. Chan, J.S. Li, E. Wong, W. Cheung, V. Ng, and W. Fok, Experimental analysis of an RFID security protocol. *The 2006 IEEE International Conference on e-Business Engineering (ICEBE 2006)*, Shanghai, China, 2006.

[26] A. Juels. Strengthening EPC Tags Against Cloning. In M. Jakobsson and R. Poovendran, eds., ACM Workshop on Wireless Security (WiSe), pp. 67–76, 2005.

[27] A. Juels, Minimalist cryptography for low-cost RFID tags. In *Proc. 4th Int. Conf. Security Commun. Netw.*, C. Blundo and S. Cimato, Eds. New York: Springer-Verlag, 2004, vol. 3352, *Lecture Notes in Computer Science*, pp. 149–164.

[28] Weis, "Security and Privacy in Radio-Frequency Identification Devices," Master's thesis, MIT, Cambridge, MA, 2003.

[29] S. Weis, S. Sarma, R. Rivest, and D. Engels, Security and privacy aspects of low-cost radio frequency identification systems, *International Conference on Security in Pervasive Computing*, March 2003, Fort Worths TX, pp. 454–469.

[30] S.M. Lee, Y.J. Hwang, D.H. Lee, and J.I. Lim, Efficient authentication for low-cost RFID systems. *International Conference on Computational Science and its Applications* (ICCSA 2005), Singapore, 2005, pp. 619–627.

[31] M. Ohkubo, K. Suzuki, and S. Kinoshita. Cryptographic approach to privacy-friendly tags. In RFID Privacy Workshop, MIT, MA, 2003.

[32] M. Ohkubo, K. Suzuki, and S. Kinoshita, Efficient hash-chain based RFID privacy protection scheme, in *Proc. Int. Conf. Ubiquitous Computing*, Nottingham, U.K., 2004.

[33] G. Avoine and P. Oechslin, A scalable and provably secure hash based RFID protocol. The *International Workshop on Pervasive Computing and Communication Security*, Kauai Island, Hawaii, 2005, pp. 110–114.

[34] E.Y. Choi, S.M. Lee, and D.H. Lee, Efficient RFID authentication protocol for ubiquitous computing environment. The *International Workshop on Security in Ubiquitous Computing Systems*, Nagasaki, Japan, 2005. Proceedings Lecture Notes in Computer Science 3823, Springer, 2005.

[35] S. Lee, T. Asano, and K. Kim, RFID mutual authentication scheme based on synchronized secret information. In *Symposium on Cryptography and Information Security*, Hiroshima, Japan, 2006.

[36] M. Ohkubo, K. Suzuki, and S. Kinoshita, Cryptographic approach to privacy friendly tags. In *RFID Privacy Workshop*, MIT, Cambridge, MA, 2003.

[37] I. Vajda, and L. Buttyn, Lightweight authentication protocols for low-cost RFID tags. *Workshop on Security in Ubiquitous Computing*, 2003.

[38] T. Dimitriou, A lightweight RFID protocol to protect against traceability and cloning attacks. *The Conference on Security and Privacy for Emerging Areas in Communication Networks*, Athens, Greece, 2005.

[39] D.N. Duc, J. Park, H. Lee, and K. Kim, Enhancing Security of EPCglobal Gen-2 RFID tag against traceability and cloning. *The 2006 Symposium on Cryptography and Information Security*, Hiroshima, Japan, 2006.

[40] D. Henrici and P. Miller, Hash-based enhancement of location privacy for radio frequency identification devices using varying identifiers. *International Workshop on Pervasive Computing and Communication Security*, Orlando, FL, 2004, pp. 149–153.

[41] D. Ranasinghe, D. Engels, and P. Cole, Security and privacy: Modest proposals for low-cost RFID systems. In *Auto-ID Labs Research Workshop*, Zurich, Switzerland, 2004.

[42] J. Katz and J.S. Shin, Parallel and concurrent security of the HB and HB+ protocols. *Advances in Cryptology—EUROCRYPT'06*, Saint Petersburg, Russia, 2006.

[43] S. Piramuthu, HB and related lightweight authentication protocols for secure RFID tag/reader authentication. In *Collaborative Electronic Commerce Technology and Research*, Basel, Switzerland, 2006.

[44] C. Castelluccia and G. Avoine. Noisy tags: A pretty good key exchange protocol for rfid tags. The 7th *IFIP Conference in Smart Card Research and Advanced Application*, Terragona, Spain, April 2006.

[45] V. Natarajan, A. Balasubramanian, S. Mishra, and R. Sridhar, Security for energy constrained RFID system. *The Fourth IEEE Workshop on Automatic Identification Advanced Technologies* (AutoID'05), Buffalo, NY, 2005, pp. 181–186.

[46] M. Lehtonen, T. Staake, F. Michahelles, and E. Fleisch. From identification to authentication—A review of RFID product authentication techniques. Workshop on RFID Security, 2006.

[47] S. Zhou, Z. Luo, E. Wong, and C.J. Tan, Interconnected RFID reader collision model and its application in reader anti-collision, *IEEE RFID Conference*, Dallas, Texas, March 2007.

[48] K. Finkenzeller. *RFID Handbook* 2nd ed. John Wiley & Sons, 195–219, 2003.

[49] D.W. Engels. The reader collision problem, AUTO-ID Center White paper, http://autoid.mit.edu/whitepapers/MIT-AUTOID-WH-007.PDF, 2002.

[50] D.W. Engels, and S.E. Sarma, The reader collision problem. *IEEE International Conference on Systems, Man and Cybernetics*, Hammamet, Tunisia, October. 2002.

[51] Y. Yuan, Z. Yang, Z. He, and J. He, Taxonomy and survey of RFID anti-collision protocols, *Computer Communications*, 29 (2006), 2150–2166.

[52] V. Deolalikar, M. Mesarina, J. Recker, D. Das, and S. Pradhan, Perturbative time and frequency allocations for RFID reader networks. in *HP Lab Technical Report*, Palo Alto, CA, 2005.

[53] J. Waldrop, D.W. Engels, and S.E. Sarma, Colorwave: An anticollision algorithm for the reader collision problem. In *IEEE International Conference on Communications (ICC'03)*, Ottawa, Canada, 2003, pp. 1206–1210.

[54] J. Waldrop, D.W. Engels, and S.E. Sanna. Colorwave: A MAC for RFID reader networks. In *IEEE Wireless Communications and Networking Conference (WCNC)*, New Orleans, LA, 2003. pp. 1701–1704.

[55] S.M. Birari and S. Iyer. PULSE: A MAC protocol for RFID Networks. *First International Workshop on RFID and Ubiquitous Sensor Networks (USN)*, Nagasaki, Japan, December 2005.

[56] S.M. Birari and S. Iyer. Mitigating the reader collision problem in RFID networks with mobile readers. In *13th IEEE International Conference on Networks*, Kuala Lumpur, Malaysia, 2005, pp. 463–468.

[57] ETSI EN 302 208-1 v1.1.1, September 2004. CTAN:http://www.etsi.org

[58] K.S. Leong, M.L. Ng, and P.H. Cole. The reader collision problem in RFID systems, *IEEE International Symposium on Microwave, Antenna, Propagation and EMC Technologies for Wireless Communications (MAPE 2005)*, pp. 658–661, Beijing, China, August 2005.

[59] S. Jain and S.R. Das. Collision avoidance in a dense RFID network. In *Proceedings of the 1st International Workshop on Wireless Network Testbeds, Experimental Evaluation and Characterization*. Los Angeles, CA, 2006, pp. 49–56.

[60] EPC Generation 1 Tag Data Standards v. 1.1 Rev.1.27, EPC Global, Tech. Rep., May 2005.

[61] EPC Radio-Frequency Identity Protocols Class-1 Generation-2 UHF RFID Protocol for Communications at 860 MHz–960 MHz V 1.0.9, EPC Global, Tech. Rep., January 2005.

[62] H. Junius, D.W. Engels and S.E. Sarma, HiQ: A hierarchical Q-learning algorithm to solve the reader collision problem, In *Proceedings of the International Symposium on Applications and the Internet Workshop*, January 2006, pp. 88–91, Phoenix, AZ, 2006.

[63] G. Tsudik, YA-TRAP: Yet another trivial RFID authentication protocol. *International Conference on Pervasive Comput. and Comm.*, Pisa, Italy, 2006.

[64] D. Gupta J. LeBrun, P. Mohapatra, and C.-N. Chuah. Reader collision avoidance mechanism in ubiquitous sensor and RFID networks. In *Proceedings of the 1st International Workshop on Wireless Network Testbeds, Experimental Evaluation and Characterization, International Conference on Mobile Computing and Networking*, Los Angeles, CA, 2006, pp. 101–102.

[65] S.-R. Lee, S.-D. Joo and C.-W. Lee, An enhanced dynamic framed slotted ALOHA algorithm for RFID tag identification. *The Second Annual International Conference on Mobile and Ubiquitous Systems: Networking and Services (MobiQuitous 2005)*. San Diego, CA, 2005, pp. 166–174.

[66] D.R. Hush and C. Wood, Analysis of tree algorithms for RFID arbitration. In *Proceedings of the IEEE International Symposium on Information Theory*, Cambridge, MA, 1998, p. 107.

[67] B. Carbunar, M.K. Ramanathan, M. Koyuturk, C. Hoffmann, and A. Grama, Redundant reader elimination in RFID systems. In *Proceedings of Sensor and Ad Hoc Communications and Networks (IEEE SECON 2005)*, Santa Clara, CA 2005, pp. 176–184.

[68] F. Wang and P. Liu. Temporal management of RFID data. In *Proceedings of the 31st International Conference on Very Large Databases*, Trondheim, Norway 2005, pp. 1128–1139.

[69] S. Chawathe, V. Krishnamurthyy, S. Ramachandrany, and S. Sarma. Managing RFID data. In *Proceedings of 30th VLDB Conference*, Toronto, Canada, 2004, pp. 1189–1195.

[70] J.D. Porter, R.E. Billo, and M.H. Mickle, Effect of active interference on the performance of radio frequency identification systems, *Int. J. Radio Frequency Identification Technology and Applications*, 1, 1: 4–23, 2006.

[71] J. Rao, S. Doraiswamy, H. Thakkar, and L.S. Colby. A deferred cleansing method for RFID data analytics. In *the Proceedings of the 32nd International Conference on Very Large Data Bases*, Seoul, Korea, 2006, pp. 175–186.

Chapter 11

Location Tracking in an Office Environment: The Nationwide Case Study

Irene Lopez de Vallejo, Stephen Hailes, Ruth Conroy-Dalton, and Alan Penn

Contents

This chapter focuses on studying organizational, social, spatial, temporal, and technical issues related to the deployment of location tracking systems in an office environment.

We have investigated and present findings on how workers perceive the experience of the surveillance and understand the scope of the deployment, the reality of wearing a tag, the pervasive nature of the organizational culture, and its influence in the attitudes observed.

We have explored these issues using an ethnographic approach and qualitative methods, participant observation, and interviews. The fieldwork has been carried out in a real office setting and has taken the form of a case study.

11.1   Introduction

Included in this chapter are the findings of a detailed case study conducted at the headquarters of Nationwide, in Swindon, United Kingdom. Participant observation, in-depth interviews, and examination of location tracking data were used to build a detailed, mainly qualitative, picture of how attitudes and behaviors are affected by the deployment of a pervasive technology. Attitudes of the workers participating in the study are explored to provide a context in which to analyze the impact of a location tracking system in the relationship between workers and the social dynamics they build through and around their physical environment.

The main findings are grouped around a number of issues:

1. Social, spatial, and temporal issues affecting the performance of the location tracking system
2. Privacy, transparency, and control: what do we fear, what can we see, and who do we think is in charge
3. The pervasiveness of the organizational culture and its mark on the acceptance of the deployment
4. Attitudes toward the technology project

The conclusions focus around recommendations for the development and deployment of indoor location-tracking systems, implications for workplace design, and the case for the development of a multidisciplinary theoretical framework. We also discuss methodological and practical limitations of the study and close the chapter highlighting possible paths of research that have come out of the current reflection.

11.2 The Case Study: Nationwide Headquarters, Swindon

Nationwide is one of the biggest financial institutions in the United Kingdom. As with many others, Nationwide is today faced with a changing and competitive market in which technology can provide an advantage; thus, technological awareness is a priority for Nationwide and other such businesses. Their motivation may range from the need to update old fashioned systems, increase performance or decrease cost, or both, to the simple desire to position themselves at the forefront of technology adopters in their area. These factors are leading organizations, such as Nationwide, to acquire and deploy highly sophisticated technologies inside their buildings; technologies exciting current interest include VoIP, Wireless, Sensor networks, and RFID (Radio Frequency*). One of the capabilities that such technologies can provide is accurate real-time information on the location of objects and individuals.

Today, facilities managers working for efficiently run organizations in big, complex buildings regularly conduct space occupation and space use studies in order to understand numbers of people in different areas of the building to be able to manage the use of space and, ultimately, to make informed decisions on investment in property. In addition, these studies contribute to planning health and safety and disaster discovery procedures.

The methods used today are manual, it being common use to employ observers who count people and note observations in templates or building plans. In public areas of buildings, the studies focus on headcounts (number of people) distributed by location (entrance, exit, hall, restaurant, ATM, etc.) and time slots. In office environments, the studies usually center on space utilization (occupation and usage of space), counting the number of people in a particular space (desk, meeting room, cafeteria area) over time.

Current approaches to acquire location data are unable to capture sufficiently rich information to give organizations the accurate picture of individuals' locations and movements across space needed to build an understanding of overall space utilization. They are also time consuming, obtrusive, and resource intensive. Organizations also want to understand spatial relationships inside the building, not only

---

* It should be acknowledged that, while this chapter adopts a primarily resource-based view of the organization, individuals and groups within organizations may have motivations at variance to the imperatives of the wider organization.

in terms of simple location and movement, but of relations between them and with the working environment.

Accurate real-time location and tracking technologies, therefore, are of particular interest to large organizations, such as Nationwide. Knowing, in real-time, the location of their workers gives companies "the option of measuring, understanding, monitoring and managing their buildings better, and the chance to investigate the relationships of the building to the day to day experience of each employee over time"* From a corporate perspective, ideally such a system would be able to locate and track individuals in space with such a precision that it is possible to get live coverage of their movements and interactions. It would record these events and allow postprocessing and analysis. Clearly the capabilities of these technologies to monitor the behavior of individual staff raises certain ethical issues.

In order to address these issues, Nationwide set up a highly ambitious and novel technology pilot project in 2005 that continued in 2006, The Smart Building Project. The project had three operational phases:

> **Phase I:** Install and test a passive RFID security system to monitor numbers of staff moving in and out the pilot area.
>
> **Phase II:** Test and apply "tags" to assets and link the individual tags to cross reference of ownership.
>
> **Phase III:** Install and test a location tracking system using active tags and ultrawide band (UWB) radio signals to monitor the position and movement of staff within the pilot area.

This case study focuses on the deployment carried out in Phase III, in which the spatial and social arrangements surrounding the introduction and deployment for six weeks of an UWB system was studied. This ultrawideband system was developed and provided by the technology supplier Ubisense, and will be referred to as the "Ubisense system."

11.2.1 Nationwide House and the Office Environment

The Ubisense system was installed in Nationwide's headquarters building, Nationwide House, a purpose built structure on the outskirts of Swindon. The building has lower ground, ground, first, and second floors. It is open 24 hours a day, 7 days a week. Its main feature as a building is a street-like layout incorporating a third of the ground floor area. This is the main public space of the building in which the shared facilities are concentrated. An Internet cafe, restaurant, convenience shop, and coffee vending machines are located along the main street area.

The pilot was conducted on one wing of the second floor, Block A (see Figure 11.1 and Figure 11.2). This space is shared by three different departments: Property

---

* Quotation from a Nationwide internal document.

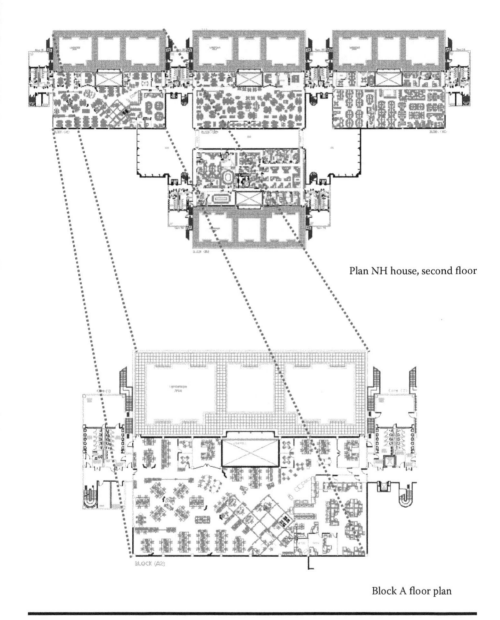

Plan NH house, second floor

Block A floor plan

Figure 11.1 Nationwide house floor plan: Block A highlighted.

Services (PS), Retail Strategy and Planning (RSP), and Legal Compliance (LC). Only the first two took part in the UWB technology pilot project with a total number of 51 people involved, which, in this case, meant carrying tags (Ubitags) during office hours.

Figure 11.2 Office environment plan and departments involved in the case study.

The three departments are accommodated in a single open plan area. The first impression is of a flowing open, although somewhat labyrinthine, space. This feeling is underlined by the use of a sinuous red carpet that runs across the length of the floor plan. The labyrinthine aspect is formed by a metallic structure that houses the PS and RSP senior management in the middle-bottom part of the plan. The mixture of openness and enclosed spaces makes for an interesting mixture of private, semiprivate, semipublic, and public areas that have different types of use, ranging from static fixed positions to highly flexible drop in, quiet and break out areas. There are two big meeting rooms that are often used by outside departments.

Corporate branding, exemplifying the corporate culture, is highly visible all around the building. The posters and signs displaying the Nationwide motto and the five-year PRIDE campaign are highly visible, as is the fact that Nationwide was voted Best Big Company to Work For in the 2005 *Sunday Times* survey.

The floor on which the pilot took place accommodated 130 staff from three different business units at the time of the pilot: PS, RSP, and LC. However, the Ubisense system was only deployed in the areas occupied by PS and RSP due to financial constraints.

The Ubisense system was deployed some weeks after the installation of a RFID-based localization system. As part of this deployment, staff from all three departments carried a credit card-like RFID tag. Before the UWB system became operational, Nationwide staff were already carrying two tags: a Nationwide security pass and the RFID pilot project tag. When the Ubisense system was deployed, the 51 staff taking part in the pilot were obliged to wear a third tag.

11.2.2 The Technology

As mentioned above, Nationwide set up the pilot project to test a number of technological possibilities in order to be able to measure, understand, monitor, and manage their buildings, how they operate, and their relationship to the day-to-day experience of each employee over time. The Ubisense system was deployed for a total period of six weeks as a temporary technology pilot. Nationwide was interested in the potential for the system to be mobile and easily deployed in different floors and buildings.

The system uses ultrawideband radio to determine locations of people and/or assets in indoor environments. Radio pulses are transmitted from tags worn by employees as they move about the office environment. The pulses are received by sensors mounted around the periphery of the building or rooms within the building and these calculate the position of the tags in real-time. The location data can be used in this raw form or it can be used to determine location events, i.e., when did a person enter the 1 m × 1 m zone in front of a desk, or how long was a person in a corridor zone. In the trial, all the data gathered was stored in a conventional SQL (structured query language) database.

The reliability of the locations and events recorded is critical. Determining location in indoor environments is well known to be very difficult to accomplish. Among the main reasons for this are:

- Radio reflections in indoor environments cause errors (multipath).
- Metal obstructions that block the direct path of the radio signal (shadowing).

The main reason for using UWB for indoor location tracking is to overcome the multipath problem.

In addition to supplying the sensor and tagging hardware, Ubisense also supplied a software application to measure space utilization. This application takes the raw XYZ data and translates it into information on the frequency of use in predefined zones in the building. The rationale for this is that such an application could provide far more reliable and useful information than can be gathered by other means (e.g., manual surveys) in a form that was useful to the end-users of the system. Ubisense predicted that it would be able to calculate the position of staff within the pilot area to an accuracy of 15 cm with a 90-percent degree of certainty.

The technology provider agreed with the organization to the following objectives:

- Install hardware and software to cover the study area and track up to 51 tags.
- Track and record the locations of employees.
- Provide services to install, monitor, and configure the system.
- Provide tailored output of the space utilization of predefined zones in the form of Excel® spreadsheets, essentially a set of analyses from the database.

In addition, a screen was installed in the pilot area showing the location and movement of the tags within the pilot area in real-time (Figure 11.3).

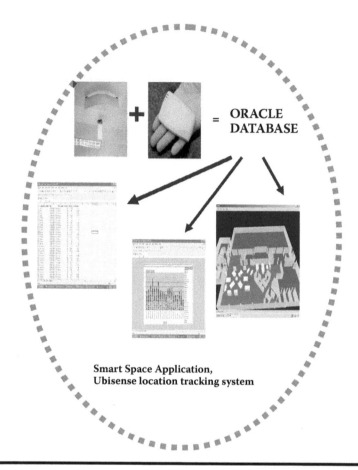

ORACLE
DATABASE

Smart Space Application,
Ubisense location tracking system

Figure 11.3 Ubisense system: Smart space application.

The setup, configuration, and initial testing to get the system into an operational state was expected to take two to three days. It was seen as highly desirable that this setup period should be as short as possible and have minimum disruption to normal working activities. In practice, this schedule proved highly optimistic. Moreover, the equipment installed proved more intrusive than anticipated. Figure 11.4 shows a sensor fixed to the ceiling. In addition, a buffer can be seen. These buffers were retrofitted to control problems with signal interference from other sensors. A problem that had not originally been anticipated.

The floor plan in Figure 11.5 shows the setup area and the sensor positions.

A total of six sensors in Area 1 were used and (ultimately) seven sensors in Area 2, significantly more than the original estimate of ten sensors to cover the area.

The timescales for the trial were roughly to start the installation, setup, and configuration on June 6, lasting two to three days, followed by approximately four

Figure 11.4 Picture of the deployment: Sensor and buffer, June 2005, Nationwide House.

weeks of data gathering, ending approximately July 1. The trial lasted six full weeks in the end.

11.2.3 Communication Strategy

A key aspect of the pilot was the extensive, well-planned effort made toward communication with the staff. Four communication channels were used: e-mail, face-to-face, visual display, and visual examples. The project champion and senior executives acted as examples of how the tag should be correctly worn. As will be seen later, wearing the tag correctly had an important impact on the performance of the system.

The strategy was intended to explain each step of the process, what was expected from staff in terms of collaboration, the importance of the project to the company, and to diffuse concerns about data protection issues.

The project champion, the head of Research and Development, gave a presentation to all staff involved in the pilot prior to its start. This presentation described

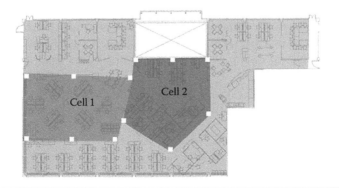

Figure 11.5 Sensor network coverage areas and sensor position.

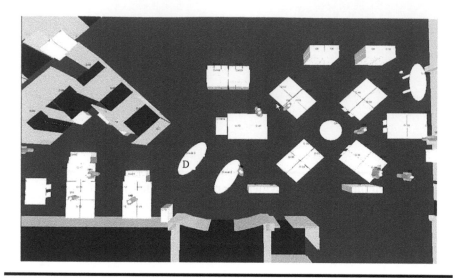

Figure 11.6   3D representation of location.

the three phases of the overall project and the technologies involved, explained that the deployments were temporary, and that participation was voluntary. One of the benefits from this presentation was to put a face and a name to the project and, thereby, open up two-way communication through e-mail and face-to-face discussions through the course of the project. The Project Champion personally handed out the two sets of tags, explaining how they should be worn, what they were for, etc., and answering the questions people posed, mainly concerning radiation fears.*

Through the presentation and e-mails, staff were informed of the physical extent of the pilot, its duration, the type of data being collected, and how the data was to be used.

The real-time display (Figure 11.6 and Figure 11.7) was provided to give a visual check on the performance of the system. While not originally intended as part of the communication strategy, by placing the real-time display in the center of the pilot area, staff had the opportunity to see and understand what data was being gathered through the pilot.

As will be seen in the conclusions, communication to staff about the pilot had a significant impact on understanding the scope of the deployment and on attitudes toward the technology.

---

* Radiation coming from either of the two tags was significantly lower than that radiated from a mobile phone.

Figure 11.7   Visual display.

11.3   Studying the Pilot

11.3.1   *Objectives*

This study should be understood as an exploratory exercise to understand the effect that location tracking technologies can have on staff in an office environment. The focus was on exploring the effects on individual and group attitudes and perceptions toward a particular technology in a particular environment. The results were expected to throw some light on organizational, spatial, and social workplace issues that affected and were affected by the deployment.

The study was carried out in three parts:

1. Participant Observation focused on space use, use of the tag, and general working behavior in the second floor office.
2. Two sets of in-depth interviews intended to understand a range of attitudes toward and around the technology.
3. A qualitative analysis of the behavior observed around the real-time visual display showing location data obtained by the UWB system and open to the whole department.

The report will be presented in four parts. Section 11.3.2 describes the methodological approach chosen, ethnography, the propositions explored, and the tools used. Section 11.4 presents the results of the participant observation and of the set of interviews organized around a number of issues that have risen from the analysis. Section 11.5 draws some conclusions and discusses the limitations of the study. The final section (11.6) introduces a series of research paths to further investigate these phenomena.

11.3.2   *Methods*

Given the aims of this study, to detail attitudes and behaviors resulting from the technology deployment, an ethnographic approach was adopted. The ethnographic approach attempts to see things from the perspective of others, to tell the story from the point of view of the user of the technology and not that of the researcher. Ethnography involves a type of observation in which the researcher is intimately involved in the social setting and the field research is a theory-generating activity. Ethnography was used as the method for observation and data collection, using in-depth interviews and participant observation as tools of data gathering and analysis.

The first set of interviews aimed at finding out, from a cross section of the department, the general experience of carrying a tag that users knew was tracking their movements; their understanding of what the location technology scope was for this group (the RFID Phase 1 "Volumes and Movement" project) in which they were already participating; their understanding of the next step; and the benefits perceived. The approach involved asking semiopen questions that allowed the informants to develop a narrative and express their opinions.

The second batch of interviews followed the same line of inquiry in order to evaluate the effect of introducing the Ubitags. These were bigger and heavier than the RFID tags, and were worn hanging from the neck instead of in a pocket or on the shirt. Moreover, specific questions were added that related to the visual display with the live location data and to the visual awareness of the deployment due to the attachment (in plain view) of the sensor network to the department's ceiling.

Interviews were conducted with individuals that were wearing the Ubitag and individuals that were not. The primary intention was to measure the impact of the UWB system deployment, but we soon realized that the results were going to be colored by the fact that this was not the first location system to be deployed in the same area. Therefore, it was decided to interview, firstly, a cross section of individuals spread across the office, across all units, including the three departments, and, secondly, a cross section of the subgroup of 51 individuals carrying the Ubitags.

A total of 28 interviews were carried out. The first set of 16 interviews were conducted over several days in June 2005, before the deployment of the Ubisystem. The second set of 12 interviews were conducted in mid-July 2005, during the last days the UWB deployment was in place. In all that follows, the names of the interviewees have been changed to preserve anonymity. Even though alias are used, the initials of the department to which they belong are included.

The interviewees can be divided into three categories. Those actively involved in the operation of the pilot (in this case only one), those belonging to a business unit anticipating to gain some benefit from the technology (the Property Services Group), and those working in the pilot area, but not members of the other two groups (in this case staff from the Retail Strategy team). As will be seen in the results, the degree of involvement in the project has proven to be an important variable in explaining attitudes toward and perceptions of the scope of the technology.

Taking into account this context, we have propositions or high-level assumptions that have guided the qualitative research along its two phases:

> **Proposition 1:** The experience of the surveillance will manifest itself in negative attitudes toward the technology deployment.
>
> **Proposition 2:** Participants in the deployment will tend to mystify the scope and capabilities of the technology.
>
> **Proposition 3:** Wearing the tag will raise complaints that will diminish through time.

We sought to eliminate experimental bias as a result of making assumptions about how these issues would evolve (positively or negatively), nor was any attempt made to predict the results or the answers to our questions. This approach facilitates a research process that unfolds and evolves rather than being prestructured (and, therefore, constrained), an important criterion given the relative lack of existing research in this area.

The propositions are used to initiate the study, but are developed as the research proceeds. Once the data has been collected, analyzed, and compared with the initial propositions, they can be revised as necessary.

11.4 Results of the Technology Pilot

The actual performance of the system in tracking staff was far poorer than predicted. Instead of the predicted 95 percent accuracy, when it was tested, the average was around 42 percent.* The disparity was due to three factors: (1) the reliability of the system, (2) the accuracy of the readings, and (3) the management and interpretation of the location data obtained.

The first two issues are related to the overall performance of the system. Its reliability was compromised due to a software problem in the basic tracking that the startup team had not encountered in previous deployments. Unfortunately, this problem was never completely resolved during the pilot. The second factor contributing to the lower than expected accuracy of the readings was the result of the prevalence of metal in the office environment. The metal influenced signal propagation, which introduced errors in locating the tags. Unfortunately, it was not possible to resolve this problem by adding additional sensors and changing sensor positions.

Finally, the difficulty of managing and interpreting the data came as an unexpected surprise for the team responsible for the deployment. Location data was filtered and input into an Excel spreadsheet, but there was no clear link between the rich granular data and the organization's need for sophisticated occupancy and utilization analysis. The development of further software applications to provide this analysis, however, expected to resolve this problem.

---

* An accuracy test was conducted at the beginning of July by Ubisense Ltd. in the office environment.

Figure 11.8 Example of the different ways of wearing the tag.

Attempts were made to overcome the accuracy problems resulting from signal interference by changing the way that users wore the tags. For the system to work effectively, the tags had to be worn in such a way as to ensure that no metal obstructed the line-of-sight between tag and sensor. While staff were quite used to carrying swipe cards to access the building, these did not need to be worn in any particular manner. For the Ubisense system tags to work, however, they had to be worn high on the body, typically hanging around the neck (Figure 11.8 and Figure 11.9). The majority of staff found this cumbersome and irritating since the Ubitags weighed some 66 g. As a result, staff frequently removed them, placing them on the desk while working and then failing to pick them up when moving around the office or, alternatively, placing them in their pockets or bags and forgetting about them.

Figure 11.9 Example of the "incentive" and communication note to encourage wearing the tag.

One of the interviewees summarized user attitudes to the tags stating, "The mechanics of actually wearing it was inconvenient. Especially on top of the fact that we've also got another extra tag at the moment anyway. So we're wandering around with three things around our necks where normally there'd be one." (**Mark PS**).

Other staff commented that "I was pretty good about wearing it" (**Anna RS**), suggesting that wearing the tags never become "normal" and was viewed as an irritation. Clearly, the temporary nature of the deployment contributed to the willingness of staff to accept the inconvenience.

Despite this, after two weeks, scarcely anyone was wearing the tags as intended: hanging from the neck and in open view, facing the ceiling and clear of any obstacles. This, combined with the fact that the desks' partitions were made out of metal, resulted in few as well as inaccurate readings.

To obtain better readings, attempts were made to change staff behavior in the third week of the deployment. This proved difficult. The project champion was able to make his staff wear the tags correctly for a few days, but, in the end, they went back to wearing it in the way that best suited their habits. When it became a nuisance, they simply dropped it. There were very few readings in the last days of the pilot as a result of this.

11.5   Results of the Study

11.5.1   *Understanding and Communication*

User understanding of the pilot was largely a result of the communication strategy. Users held conflicting perceptions on the success of this strategy. A slight majority of the staff believed that the communication strategy had been successful. "I actually thought the communication was quite good and that it was communicated clearly" (**Gwen PS**). On the other hand, a significant minority were critical, complaining about a lack of communication.

The informants' understanding of the deployment in terms of objectives proved fairly good and there was a common agreement on the terms used to describe it: *movement, space utilization, granular, granularity, workspace utilization, better working environment.* This is a reflection of the internal communication campaign accompanying the project that has helped to develop a shared understanding of the pilot aims. Understanding was not, however, uniform. "It is difficult to know what you are trying to get out of the project" (**Anna RS**).

Despite the efforts put into communication, staff failed to understand the project scope and physical extent of the pilot. None of the respondents, apart from individuals working very closely on the deployment, could explain the scope of the tracking: in which areas it was happening and in which it was not. When asked about boundaries, a typical response was: "I have got no idea. I don't know what area is actually being measured" (**Shaun PS**). And this despite the fact the sensor network was hanging from the ceiling in open view with the sensor boxes pointing inwards.

Users generally did not understand the purpose of the data and they did not understand how the raw data was being used to analyze space utilization. Although the communication process was used to diffuse concerns about data protection, this aspect of the pilot was not covered in the communication plan. As we observe in section 11.5.3 on privacy, misconceptions about the way data was being used did lead to concerns among staff.

11.5.2   The Temporary Nature of the Deployment

Acceptance of the deployment was significantly eased by the fact that the deployment was just a trial and was temporary in nature. "I had no real problem with it. I understand this was just a trial, wasn't it?" (**Carol RS**). The fact that the deployment was temporary meant that it was not perceived as genuine. "So, you know, if you did it for real"(**Robert PS**). This fact hugely influences concerns about privacy and other attitudes toward the technology.

11.5.3   Privacy and Intrusion

The real-time display in the center of the pilot area provided an element of openness regarding the data being collected by the system, allowing the staff to see exactly what data was being gathered. Staff did, however, have some concerns about privacy, despite the communication plan and the real-time display.

Participants reacted to the display in two ways, one set of respondents looked at it, found it interesting and looked for themselves, but could not make out where they were and if they were moving, and, therefore, lost interest. This group also found it open, a feature that showed there was "nothing to hide" (**Gwen PS**). The other set did not look at it, lacked interest in it, and had a general feeling that it was "a PR exercise" (**Matthew RS**).

Staff concerns about privacy appear to be strongly influenced by their perceptions of the system's capabilities and operation. Despite the efforts put into communication, the majority of the beliefs of staff regarding the system were at variance with the reality.

With regard to privacy, there is a difference between location data being gathered by a system and the system providing the capability to extract information about individual's behavior. The Ubisense system did not provide this second capability, but staff did have concerns about their privacy being invaded by being observed all of the time at work.

There was a high degree of uniformity among participants in their misunderstanding of this aspect of the pilot. None of the respondents, apart from those working very closely on the deployment, could even explain the scope of the tracking, in which areas it was happening, and in which it was not.

The uniformity of staff perceptions regarding how they were being tracked could be understood using the concept of "collective imaginaries"* built by the informants around the technology deployment. As a result of the experience of the deployment, informants developed a thought structure that combined rumor and reason that resulted in the mystification of the spatial and technical scope of the technology and the manipulation of the location data obtained.

These collective imaginaries are constructed within social groups exposed to the same experience and information and takes the form of a shared belief. Discussions and conversations between group members may serve only to strengthen and entrench this shared impression. With regard to the staff at Nationwide, a majority thought that they were being tracked around the whole building, into the toilets, and that "someone" was able to know exactly what they were doing at all times and with whom. It is plausible that the collective imaginaries that developed were influenced by the national context, with considerable media attention devoted to debates on identity cards, CCTV, and, perhaps in part, by reality TV shows, such as *Big Brother*. The shared organizational context and culture also is likely to have had an influence.

Again, it should be observed that not all staff shared the collective belief regarding the extent to which the system was being used to monitor individual behavior. However, it may be pertinent to note that the only staff member explicitly to reject this concern (stating, "It's not being used to track you around the building" (Laura PS), worked in the Property Services group with a responsibility for security.

11.5.4 *Attitudes and Organizational Culture*

As noted earlier, Nationwide places considerable emphasis on the creation of a common organizational culture. This is reflected in the strong branding around the building. Large numbers of images and messages are displayed proclaiming the organizations values, the degree to which they care about their staff, advertising the exhibits they organize every week, the superb canteen, the Starbucks coffee, etc.

The interview responses support the fact that there is a strong organizational culture and suggest that it had a positive impact on staff attitudes to the deployment.

Key aspects of the influence of the culture relate to trust and to pride in the organization. The general perception that Nationwide is a fair and open organization appears to have been transferred into attitudes to the deployment. Staff explicitly

---

* The concept of collective imaginary refers to the body of symbolic landmarks through which any collectivity inserts itself in time and space. This process involves the establishment of four relationships:

- A relationship to space, resulting in a territoriality.
- A relationship to Self and to Others, giving shape to an identity.
- A relationship to the past, which is expressed in a collective memory.
- A relationship to the future, which is expressed in utopias: http://www.uqac.ca/_bouchard/chaire_desc_e.html.

cited their trust in the organization to explain their lack of concern of the potential for the technology to be used to monitor them. Responses to questions about the use of the data from the pilot, such as "No, I think Nationwide is a fair organization" (**Andrew PS**), were typical.

Responses also suggest pride in being part of an advanced, forward-thinking company. The deployment of an advanced technology could be seen to fit with staff perceptions about their organization and in itself become a focus for pride in the company. "There doesn't seem to be anybody else sort of doing this work… this kind of smart work environment that we've got here because we're kind of the first department to have this environment" (**Shaun PS**), and "Nationwide sets themselves as a, you know, benchmark for other people, to come in and have a look at what we're doing. So we're at the forefront, which is good" (**Albert PS**).

While a strong organizational culture exists in Nationwide, it would be wrong to depict it as all pervading or perhaps totalitarian. The organizational culture is for most, if not all staff, only one of a number of cultures within which they are immersed. Their response and commitment to the organizational culture is mediated by the influence of national, local, religious, and, potentially, class cultures.

Underneath the widespread acceptance of the organizational values and ethos, however, a current of cynicism was apparent, suggesting that the overt organizational culture, while powerful, is not ubiquitous. Surreptitious reluctance and even bitterness for making participation in the project ostensibly voluntary while making it clear that those who did not take part were acting against the organizational modus operandi, ran strongly in some interviews.

Dissention from the accepted ethos was, however, rarely explicitly stated, rather being conveyed through tone of voice, expression, and body language. Clearly no organization such as Nationwide can ever maintain nor, in all likelihood, would it seek to maintain, an all pervading influence on the views and perceptions of individual staff.

11.6 Conclusions

11.6.1 Conclusions Regarding the Pilot

The results of the pilot were mixed. Installation was more time consuming and intrusive than anticipated and the system never delivered the promised degree of accuracy.

These problems were caused by two related issues.

1. The pervasive presence of metal in the environment interfered with the transmission of the signals.
2. Users' preferences on how tags were actually worn and carried conflicted with the operational requirements of the system.

Metal partitioning is extremely common in office environments and its presence should have been foreseen. Developing technologies away from the real world context in which they will be deployed is likely to result in these kinds of problems.

Users' habits and ingrained behaviors are extremely hard to change. Technologies that require these changes are unlikely to succeed, in the absence of a strong sense of perceived benefit. However, the data gathered by the system can provide clear insights into how buildings are actually used through time and offers real potential for improving our understanding of human behavior in office environments. If these potential benefits are to be realized, more sophisticated applications will be required to translate the raw XYZ data into useful information.

Studying the deployment of a location system in an office environment has brought to light the "perfect system versus degraded environment dilemma". There is a fundamental difference, and different results too, between setting up a technology experiment in a controlled environment and taking it into real life. As we have observed, real environments are complex and unstable for a combination of spatial and social reasons, among others. The majority of computer scientists tend to think about "perfect systems" that work very well in a "perfect environment" in which those two variables are controlled. When faced with a real environment, this is perceived as degraded because it does not provide the same controlled and perfect space. In this case, the technology provider had made assumptions, given the performance of the system in controlled environments they knew well, such as their own office in Cambridge, that do not apply to all buildings or organizations. Consequently, expectation mismatches and disappointment can follow the deployment of technology.

It is our contention, as a result of the experiments conducted at Nationwide, that this approach is naive. Real deployments, and, in particular, the engagement of real stakeholders, should become requirements of the test environment in which the claims made for such systems are validated.

11.6.2   Conclusions Regarding the Study of the Pilot

Given the efforts put into communicating to staff, it is perhaps surprising to discover how poorly the staff understood the pilot. This lack of understanding was not, however, uniform. Unsurprisingly, perhaps, those most closely involved had the most accurate understanding. Perhaps more significantly, staff from Property Services, who had direct involvement in the pilot, generally had a better understanding of its objectives, scope, and extent. This may be due to closer informal contact with staff who were directly involved and, hence, had a better understanding. It also may be due, in part, to the fact that it was widely understood that this pilot would benefit their group, even if the nature of the benefit was poorly understood.

The relative failure of the communication strategy can be explained by a number of factors. While presentations were given to staff and information was transmitted through e-mails and other means, these communications were competing for the limited attention busy staff have with other communications directly relating to their jobs and roles. In the face of this competition, staff with no attachment to the pilot simply do not appear to have taken the time to read and understand the communications sent to them.

A recommendation would be to have more face-to-face communication, informal talks, and information sessions, in small groups with a proactive approach, in order to obtain the engagement of the participants. In the case reported here, only 1 out of 15 interviewees was present at the presentation given at the beginning of the project.

Lack of attachment or sense of involvement can largely be explained by the fact that the system provided no tangible benefits to staff outside the Property Services group. The ability of these systems to offer tangible benefits to all users is likely to be an important factor in staff's understanding and ultimately acceptance of the location tracking systems. A successful invasive office technology must provide tangible and immediate benefits to the workers involved. That holds for all types of systems, whether collaborative or not, synchronous or asynchronous.

Thus, the deployment of a location tracking system in an office environment should provide a set of benefits and such benefits are usually perceived through functionalities of a system and their direct impact on the personal working experience of an individual. In this pilot only one such direct functionality existed—the stand alone computer acting as a 3D visual display of the real-time tracking movements. Nobody saw this as a benefit.

Belonging is an influential variable on the perception of benefits. Moreover, given the personal sensitivities involved, good communication is key to understanding the data obtained and allowing for its effective interpretation and manipulation.

Recommendations for the design of a deployment should take into account these findings and adopt a multilevel deployment strategy involving at a minimum the social, spatial, technological, and temporal issues discussed within this chapter. Issues of privacy, transparency, and control are influenced by collective imaginaries that mystify the technology and increase fears of privacy invasion, lack of transparency over the data, and losing control over the manipulation and post processing of the data. The influence diminishes, and, therefore, objective understanding of the deployment improves if one is involved in the project and trust in the organization is solid. It would be a mistake, however, to assume that, outside the protective environment afforded by the company, one would see the same reactions to privacy-invading measures of the individuals involved.

11.6.3 Conclusions Regarding the Approach to the Study of the Pilot

The choice of an exploratory research approach and of ethnography as our fieldwork approach can be criticized. Critics point out the nongeneralizable nature of this approach, the fact that is not "representative, and therefore we cannot conclude that characteristics of the sample will be found in the larger population". We argue that any findings generated by this kind of research should be regarded as a stepping stone to future studies, exactly as we intended.

The third issue is linked to the choice of ethnography as a fieldwork approach. Critiques highlight its journalistic, unscientific, descriptive, nonanalytical, and subjective nature. We say that it is a style of research that is committed to studying people's understandings, meanings, and practices, and that fits very well with our research propositions. Our primary intent was to give voice to people's reactions and understanding of the technology.

For this particular case, an exploratory approach was necessary because of the lack of similar previous research in the field. It proved to be a choice that, coupled with an ethnographic approach, provided a fruitful approach to the study of pervasive technologies in the office environment.

11.7 Open Issues

The richness of the raw data is barely harnessed by the applications in the current system. There is a challenge for software developers in partnership with workplace designers, facilities management specialists, and social scientists to create further applications that can exploit the richness of this data.

These applications could include applications for social/spatial data mining with improved visualizations of these relationships.

A further research challenge is for social scientists and technologists to take existing theories on how technology deployment is influenced by social issues and develop them to include the spatial dimension.

The information derived from these location tracking technologies has considerable potential to inform workplace design. This presents a challenge to workplace design professionals to develop evidence-based approaches to workplace design.

The ultimate challenge would be for technology developers to use the insights from social scientists regarding the user responses to the deployment within their approach to technology development. The suggestion is that a more user-centered development approach, combined with early pilots in real environments, could lead to technologies more likely to gain user acceptance.

Acknowledgments

The authors would like to acknowledge the financial support provided by the EC-funded RUNES project, and to express our sincere appreciation to the Nationwide building society and, in particular, Fred J. Child, for allowing us to conduct this study. Child, Pete Steggles, and David Theriault, from Ubisense Ltd., have read this chapter and provided comments. We would further like to thank Paul Wheeler from Land Securities Trillium for constructive discussions over the course of the study.

Bibliography

Aipperspach, R. et al., Maps of our lives: Sensing people and objects together in the home, Electrical Engineering and Computer Sciences University of California at Berkeley, Technical report No. UCB/EECS-2005-22, November 30, 2005.

Anderson, B., *Imagined Communities*. Verso, London, 1991.

Appadurai, A., *Modernity at Large*. University of Minnesota Press, Minneapolis, 1996.

Babbie, E., *The Practice of Social Research*, 10th ed. Thomson Wadsworth, Belmont, CA, 2004.

Barney, J.B., Fuerst, B., and Mata, F., Information technology and sustained competitive advantage: A resource-based analysis, *MIS Quarterly*, 19, 1996, pp. 487–505.

Barney, J.B., The resource-based theory of the firm, *Organization Science*, 7, 5, 1996, pp. 469.

Castoriadis, C., *The Imaginary Institution of Society*. MIT University Press, Cambridge, MA 1987.

Child, F.J., Capacity Management Pilot Project for "Proof of Concept:" Business Requirements, Nationwide Internal document, February 2005.

Child, F.J., RFID Pilot Project Update to A2 teams, Nationwide Internal document, May 2005.

Harrison, A. and Steggles, P., Evaluation of an AmI workplace, in the Proceedings of e-Challenge 2004 International Conference, Vienna, Austria, October 2004.

Hatch, M.J., *Organization Theory: Modern Symbolic and Postmodern Perspectives*, Oxford University Press, Oxford, U.K., 1997.

Hillier, B. and Hanson, J., *The Social Logic of Space*, Cambridge University Press, London, 1989.

Hillier, B., *Space Is the Machine*, Cambridge University Press, London 1996.

Lopez de Vallejo, I. et al., Ambient intelligence in manufacturing: organizational implications. Workshop on *Ubiquitous Computing and Effects on Social Issues' CHI05*, Portland, OR, April 2005.

Lopez de Vallejo, I., Ambient intelligence vision: Exploring the social risks of its construction. In *The Proceedings of e-Challenges 2004 International Conference*, Vienna, Austria. October 2004.

Lopez de Vallejo, I., Soft factors in the new ICT-powered workspace. In *The Proceedings of e-Challenges 2003 International Conference*, Bologna, Italy, October 2003.

Mannings, R., Whereness: Ubiquitous positioning, *Journal of the Communication Network*, 4, I, January–March 2005.

Mills, C.W., *The Sociological Imagination*, Oxford University Press, New York, 1959.

Norman, D.A., *The Invisible Computing*, MIT Press, Cambridge, MA, 1998.

Orlikowski, W., The duality of technology: Rethinking the concept of technology in organizations, *Organization Science*, 3, 3, Focused Issue: *Management of Technology*, August 1992, p. 405.

Parliamentary Office of Science and Technology, POST note 263, Pervasive Computing, London, May 2006.

Penn, A., Space for innovation: effects of spatial configuration on social and knowledge generation. In *Proceedings of the First Workshop for Cooperation between Japan and the United Kingdom on SOFT Science and Technology*, 126-8 and 172-6, November 1994, STA, Osaka, Japan.

Penn, A., Desyllas, J., and Vaughan, L., The space of innovation: Interaction and communication in the work environment, *Environment and Planning B: Planning and Design*, 1999, vol. 26, 193–218, Pion Publishing, London.

Seale, C., *Social Research Methods: A Reader*, Routledge, New York, 2003.

Spiliopoulou, G., and Penn, A., Organisations as multi-layer networks: face to face, email and telephone interaction in the workplace. In the *Proceedings of the Second International Space Syntax Symposium*, March 1999, Brasilia, Bartlett, UCL, London, 1999, vol. 1, p. 13.

Spradley, J., and McCurdy, D., *The Cultural Experience*, Waveland Press, Prospect Heights, IL, 1972.

Spradley, James P., ed. *Participant Observation*, Holt, Rinehart and Winston, New York, 1979.

Spradley, James P., ed. *The Ethnographic Interview*, Holt, Rinehart and Winston, New York, 1980.

Steggles, P. and Cadman, J., A Comparison of RF Tag Location Products for Real World Applications: A Ubisense White Paper, Ubisense, Cambridge, U.K., March 2004.

Taylor, C., *Modern Social Imaginaries*. Duke University Press, Durham, VC, 2004.

Ubisense Web site: http://www.ubisense.net

Ubisense, Ltd., *Summary Report on Evaluation Trial at Nationwide*, June 2005, prepared for Nationwide Ltd., Nationwide Internal document, 27 July 2005. (www.ubisense.com).

Ward, A., Webster, P., and Batty, P., *Local Positioning Systems: Technology Overview and Applications*, Ubisense, Cambridge, U.K., September 2003.

Ward, A. and Cadman, J., Deploying commercial location aware systems, *UbiComp 2003*, Seattle, WA, October 2003.

Weiser, M ,The computer of the 21st century, *Scientific American*, September 1991.

Weiser, M., Some computer science issues in ubiquitous computing, *Communications of the ACM*, July 1993.

Weiser, M. and Seely Brown, J., The technologist's responsibilities and social change, *Computer Mediated Communication Magazine*, 2, 4, April 1995.

Weiser, M. and Seely Brown, J., The coming age of calm technology, *PowerGrid Journal*, v. 1.01, http://powergrid.electricit.com/1.01 (July 1996).

Chapter 12

Pervasive Computing Security: Bluetooth® Example

Giorgos Kostopoulos, Paris Kitsos,
and Odysseas Koufopavlou

Contents

12.1   Introduction

The proliferation of wireless handheld devices and advances in network technologies, and the desire to stay connected anytime and anywhere, has ushered in the domain of mobile and pervasive computing. Wireless mobile phones that also serve as personal digital assistants (PDAs) will become the norm in everyday life. People carrying these devices will be surrounded by intelligent and networked devices and embedded computing appliances, e.g., at home and work, in our cars, in public spaces, etc. The fact that we can use them everywhere drastically changes the way we use computers and imposes new requirements on computing and communications as well in applications, in general.

Mobile computing targets the mobility of users, data and systems, information organizations, and access methods for data services and management in mobile systems. Pervasive computing builds on mobile computing techniques and concerns the creation of environments populated with human beings, wireless networking devices, and physical objects that satisfy the properties of physical integration and instantaneous interoperation. In pervasive computing, the actual terminals can be hidden from the user or can be disguised as something else (like a coffee cup). Pervasive seems to imply also some kind of distribution of computation (i.e., "chip rich" environments) among different devices. Bluetooth® [1] is a technology and standard that is designed as a wireless cable replacement, which connects a wide range of devices. In contrast to wireless LANs, (local area networks) such as 802.11b, it was designed to be low powered, operate over a sort range, and support both data and voice services. Bluetooth enables peer-to-peer communications among all sorts of devices. This system enables computing and communication devices to exchange information and work together for the benefit of the user. Bluetooth and pervasive computing have much in common: Both aim to make computing and communications easier, more convenient, and more personal. Both enable the use of a myriad of devices, especially small mobile devices, to accomplish these objectives. Thus, it appears evident that Bluetooth can be an excellent match with and a key element of pervasive computing. Specifically, Bluetooth has applications in at least three important pervasive computing domains: home networking, automobile network solutions, and mobile E-business (Figure 12.1).

The Bluetooth standard aims to achieve interconnectivity between any Bluetooth device regardless of brand or manufacturer. As a result, any Bluetooth device anywhere in the world can connect to other similar devices within its range.

There are many other Bluetooth usages, such as pervasive computing. Like wireless Ethernet, Bluetooth uses radio frequency-based data transmission. The major difference between Bluetooth and wireless Ethernet is the concept of the piconet. Bluetooth is based on the idea that devices that are within close range should establish a small network, allowing them to access each other's resources. In contrast to wireless Ethernet, which is based on more or less static IP (Internet Protocol) addresses, Bluetooth enables a less restrictive, low-level communication; also, the usage of such as Bluetooth-enabled active tags referred to as BTnodes [2]. The main reason for using Bluetooth as a communication standard for the active tags is that Bluetooth modules are being integrated in an increasing number of consumer devices, such as mobile phones, PDAs, and digital cameras.

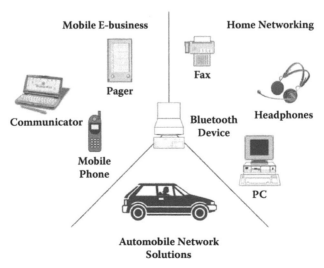

Figure 12.1 Bluetooth applications.

This system enables computing and communications devices to exchange information and to work together for the benefit of the user. Bluetooth operates in the 2.45 GHz ISM (Industrial, Scientific, Medical) band and can be used all over the world. The communications between Bluetooth devices will hop between 1,600 different frequencies per second across a total of 79 frequencies, moving in a quasi-random fashion. Air interface traffic will also be encrypted using a key between 8 and 128 bits. Applications using Bluetooth can provide their own encryption layer. This multilayered approach considerably reduces the possibility of undetected and undesirable intrusion [3].

The operating system designers and the computers' administrators had to think of ways to restrict users from reading confidential data, destroying data that did not belong to them, and doing countless other acts, which could compromise the security of the system. Networking created new problems: eavesdropping on transmitted data and malicious computers on the networks. As corporations began to use the networks, more and more financially important data was being transferred. Portable handheld and embedded devices have severely limited processing power, memory capacities, software support, and bandwidth characteristics. Also, hardware and software environments are becoming increasingly heterogeneous, a trend that will continue in the foreseeable future. Finally, security information in different domains is subject to inconsistent interpretations in such an open, distributed environment.

12.2 Bluetooth Security Layer Specification Description

Bluetooth includes security features at the link level. It supports authentication, encryption, and key management. These features are based on a secret link key that is shared by a pair of devices. To generate this key, a pairing procedure is used when the two devices communicate for the first time.

The link level functions are defined in the Bluetooth Baseband and the Link Manager Protocol (LMP) Specifications. The generic access profile specifies three security modes for a device. In security mode 1, the nonsecure mode, a device will not initiate any security procedure. In security mode 2, the service level-enforced security mode, a device does not initiate security procedures before channel establishment at the L2CAP level. This mode allows flexible access policies for applications, especially when applications with different security requirements are run in parallel. In security mode 3, the link level-enforced security mode, a device initiates security procedures before the link setup at the LMP level is completed. According to the Bluetooth security level for the device, security level 1 uses untrusted devices and level 2 uses trusted devices only. As the Bluetooth security level for service, security level 1 requires authorization and authentication, security level 2 requires authentication only, and level 3 opens all devices [1].

Bluetooth layer security uses four key elements: a Bluetooth device address, two separate key types (authentication and encryption), and a random number. The authentication key generates the encryption key during the authentication phase. A random or pseudo-random process is used in order to produce the random number. The security layer includes key management, generating key mechanisms, user and device authentication, and data encryption. The Bluetooth security architecture can secure all the upper layers by enforcing the Bluetooth security policy, or it can be flexible enough to let those layers use their own security policies.

12.2.1 Key Management

When a connection starts, the parties need to enter their personal identification numbers (PINs) (Figure 12.2). By using this PIN, the link key generation function (E_2) generates the appropriate link key. This key is used by the authentication function (E_1) in order to authenticate the identity of the connection parties. In case that data encryption is required, the encryption key generation function (E_3) produces the encryption key by using the same link key. The encryption key is used by the E_0 encryption function in order to encrypt or decrypt the sending or the receiving data.

During a connection several different types of 128-bit link keys are produced. These keys handle all security transactions between two or more parties. Depending on the type of application, the link keys can be initialization or unit, or combination or master keys. The lifetime of a link key depends on whether it is a semipermanent

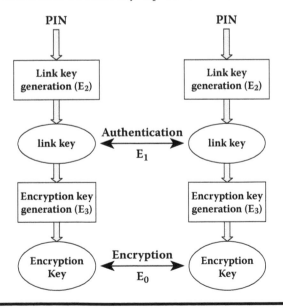

Figure 12.2   Key generation process.

or a temporary key. In a point-to-multipoint connection, where the same information is transmitted to several recipients, a master (temporary) key is used. All the other types of link keys are semipermanent. In addition to the link keys, there is the encryption key. The encryption key generation process uses the current link key in order to produce the encryption key.

In the initialization connection phase between two parties, each party produces its own initialization key K_{INIT}. This key is generated by the key generation function (E_{22}). The input parameters of the E_{22} are the PIN code, the Bluetooth device address, and a 128-bit random number. The initialization key K_{INIT} is used in order to produce the next link keys.

The two parties decide which link key (combination or unit key) will produce. In case a device has limited memory, it chooses to produce a unit key. In this case, the device does not need to remember different keys for different links. The key generation function (E_{21}) generates the unit key. The input parameters of the E_{21} are the K_{INIT}, the Bluetooth device address, and a 128-bit random number. The two parties may also decide to use the combination key. In this case, in both devices, the key generation function E_{21} generates temporary keys. These keys are exchanged between the two devices and are used in order to produce the combination keys.

When a master device broadcasts data to several slave devices, it uses the master key. The master device generates the master key in the key generation function E_{22} by using two 128-bit random numbers. Then, the master device sends to all the slave devices a random number. In all the devices (master and slaves), the generation function (E_{22}) uses this random number and the current link key in order to produce the overlay (OVL) number. Then, the master device sends to all the slaves the master key bitwise XORed with the overlay. So, each slave can calculate the master key.

12.2.2 Authentication

The Bluetooth authentication uses a challenge-response scheme where it is used to check whether the other party knows the link key. Authentication is accomplished by using the current link key. The process is symmetric, so a successful authentication is based on the fact that both parties share the same key.

First, the verifier (master) sends to the claimant (slave) a random number A (Figure 12.3). Then, both parties use the authentication function (E_1), with input of the random number A, the claimants' address (Bluetooth device address B), and the current link key in order to produce the signed response (SRES). Number L denotes the length of the actual current link key. After this, the claimant sends the SRES to the verifier. Finally, the verifier checks the matching of the two SRESs.

The verifier may not necessarily be the master. Some of the applications require only one-way authentication. In both parties, E_1 computes and stores the authenticated ciphering offset (ACO), which is used for the encryption key generation.

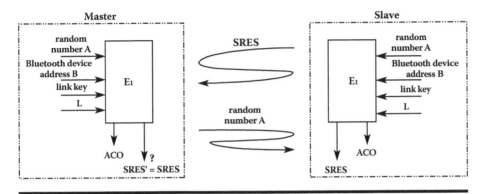

Figure 12.3   Authentication process.

12.2.3   Encryption

When the Link Manager activates the encryption, the encryption key is generated. The key generation function (E_3) generates the encryption key K_c, from the current link key, a 96-bit ciphering offset number (COF) and a 128-bit random number. TheCOF is based on the ACO, which is generated during the authentication process. The encryption key size varies from 8 to 128 bit. In a Bluetooth system, only the payload of the packet is encrypted. This is achieved by using the stream cipher function (E_0). The input of the E_0 are the encryption key K_c, the Bluetooth device address A, and some data of the master real-time clock (Figure 12.4). E_0 consists of the payload key generator, the key stream generator, and the encryption/decryption part [1].

The payload key generator combines the E_0 input bits in an appropriate order and leads them to the four linear feedback shift registers (LSFR) of the key stream generator [4]. Different encryption modes are used depending on the party link key type. The broadcast traffic is not encrypted in the case that a combination or unit

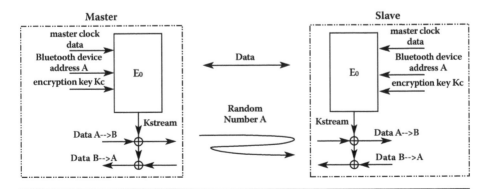

Figure 12.4   Encryption process.

key is used. If a master key is used, there are three possible encryption modes. In encryption mode 1, nothing is encrypted. In encryption mode 2, broadcast traffic is not encrypted, but the individually addressed traffic is encrypted with the master key. Finally, in encryption mode 3, all traffic is encrypted with the master key.

12.3 Bluetooth Security Layer Architectures and Implementations

Until now there are many designing approaches for Bluetooth depending on the kind of the application. For example, some companies (e.g., Philips, Ericsson, Zeevo) have been integrating the basic Bluetooth baseband functions in hardware. Other companies, like Oki Semiconductor and Intel, provide an HCI (host control interface) that can be controlled, for example, across a USB (Universal Serial Bus) interface. In [5], an analytical description of the hardware architecture and implementation of the Bluetooth security is analyzed, dictated by the official specifications. Also, in [6] an implementation of a new enhanced security layer (ESL) is presented. The security level is increased by replacing the encryption with AES (Advanced Encryption Standard) and adding integrity protection. The security processing is implemented in hardware for high performance. Finally, in [7] a partitioning hardware and software implementation is introduced. Also, the resource-sharing scheme is adopted. The partitioning and resource-sharing reduce hardware resources though the timing requirement is fulfilled.

12.4 Hardware Implementation of the Bluetooth Security Specifications

The proposed system architecture for the Bluetooth security hardware implementation is depicted in Figure 12.5. It consists of six units: (1) E_2 link key generation function, (2) E_1/E_3 function, (3) RAM/ROM memories, (4) E_0 stream cipher function, (5) control, and (6) I/O (input/output) Interface.

The E_2 link key generation function unit produces the appropriate link keys. By using these link keys, the E_1/E_3 function unit performs the authentication and produces the encryption key. For data encryption, the E_0 stream cipher function unit is used. All the interunit data movements are being transferred through the 64-bit data bus. RAM memory has also been integrated in order to store the appropriate function keys. All the function constants are stored in the ROM memory. The control unit synchronizes the operation of the system. The system communicates with the external environment through the I/O interface. The basic component of all the key generation functions (E_2, E_3) and authentication function (E_1) is the SAFER+ block cipher [1].

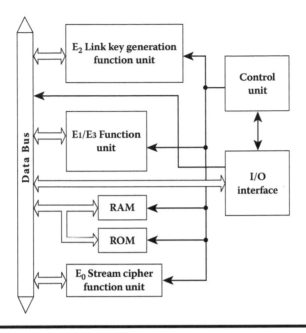

Figure 12.5   Bluetooth security system architecture.

12.4.1 *SAFER+ Block Cipher Hardware Architecture*

The hardware implementation of the SAFER+ block cipher is illustrated in Figure 12.6. It consists of two main units: the encryption data path and the key-scheduling unit. The key-scheduling unit allows an on-the-fly computation of the round keys. In order to reduce the covered area, eight loops of a key scheduling, single round implementation is used. The bias vectors are implemented by using 17×16-byte ROM blocks.

Round keys are applied in parallel in the encryption data path. For the full SAFER+ algorithm execution, eight loops of the single round are needed. The single round hardware implementation solution was chosen because, with this minimum resource covered area, the required throughput can be achieved.

The first component of the encryption data path is an Input register, which concatenates the Plaintext and the Feedback data produced in the previous round. The input register feeds the SAFER+ single round. After the SAFER+ single round in the encryption data path is a mixed addition/XOR component, which combines the feedback data with round subkey K_{17}. The final stage of the encryption data path is the Output register.

The four subunits of a SAFER+ single round (Figure 12.7) are:

1. The mixed XOR/addition subunit, which combines data with appropriate round subkey K_{2r-1}.

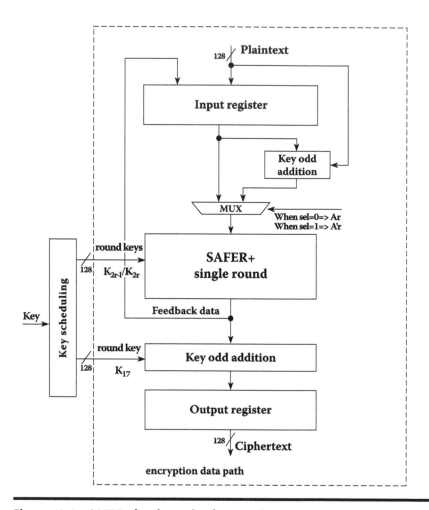

Figure 12.6 SAFER+ hardware implementation.

2. The nonlinear layer (usage of the nonlinear functions "e" and "l"). The "e" function is implemented as $y = 45^x$ in GF(257) with exception that $45^{128} = 0$, and "l" function is implemented as $y = \log_{45}(x)$ in GF(257) with exception that $\log_{45}(0) = 128$.

3. The mixed addition/XOR subunit, which combines data with round subkey K_{2r}.

4. The four linear PHT layers connected through an "Armenian Shuffle" permutation.

The nonlinear layer is implemented using a data mapping component that produces the X1 and X2 bytes. These bytes are the input of the nonlinear functions "e" and "l". During one round, both functions, "e" and "l", are executed eight times.

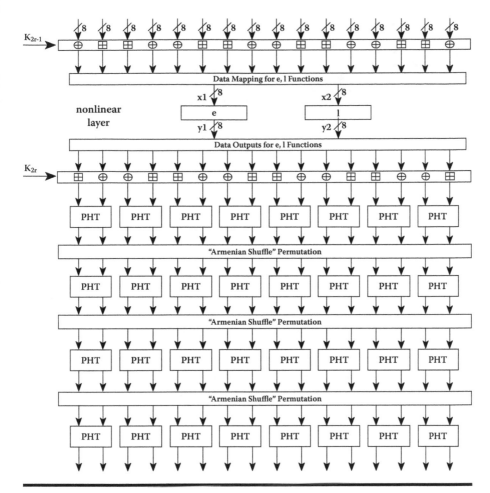

Figure 12.7 SAFER+ single round.

With this design, a significant reduction of the required covered area is achieved. Each function is implemented by using 256 bytes ROM memory.

In the following, the SAFER+ block cipher will be denoted as function Ar. The E_1 function for the authentication procedure uses the encryption functions Ar and A'r. There is only one difference between the two functions. In A'r, the input of the first round is added to the input of the third round. The Ar and A'r functions are implemented as illustrated in Figure 12.6. The signal "sel" determines the selection between Ar or A'r.

12.4.2 *The Link Key Generation Function E_2*

The architecture of the function, E_2, which is responsible for the link key generation, is shown in Figure 12.8. Both E_{21} and E_{22} key generation functions are used

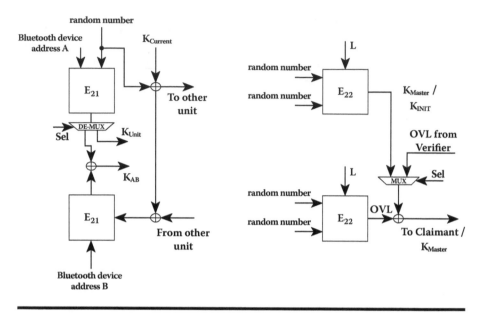

Figure 12.8 E$_2$ function architecture: unit-, combination-, initialization-, and master key production.

for the creation of the four authentication keys. These functions are implemented by using mainly the SAFER+ algorithm. The four keys are grouped in pairs. The unit key is paired with the combination key, and the initialization key is paired with the master key. The keys after generation are stored in a 5 × 128-bit RAM memory.

12.4.3 The E$_1$/E$_3$ Function Architecture

The architecture of the E$_1$ and E$_3$ functions are similar, so these architectures were implemented in a reconfigurable unit. The only difference is in their expansion subunit. The expansion subunit expands either the Bluetooth device address or the COF parameter into a 128-bit vector. The offset module produces the key for the A'r function. The Ar function needs 16 cycles in order for a new set of SRES and ACO parameters to be produced. If encryption is needed, the Link Manager sets the encryption mode signal, so the unit operates as E$_3$ function. After 16 cycles, the E$_3$ function creates the encryption key K$_c$. A detailed diagram of the E$_1$/E$_3$ architecture is illustrated in Figure 12.9.

12.4.4 The E$_0$ Unit Architecture

Two components are needed for the implementation of the E$_0$ stream cipher function unit (Figure 12.10). The first is the payload key generator component, which

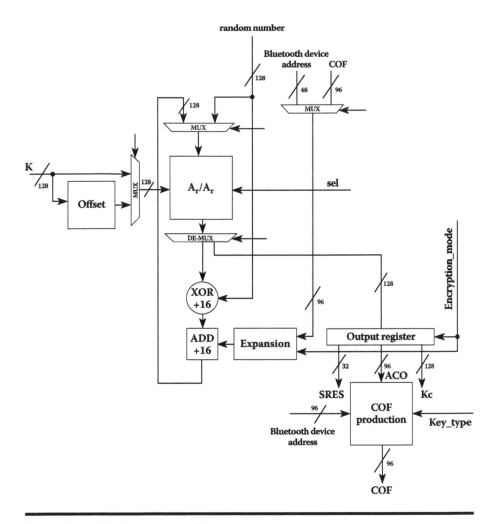

Figure 12.9   The E_1/E_3 unit architecture.

has as input the encryption key K_c, the 48-bit Bluetooth device address, and the 26-bit master clock data. The second component (key stream generator) generates the key stream. It uses four LFSR's. The LFSRs output is the input of a 16-state finite state machine. The state machine output is the key stream sequence. In the payload key generator component, the encryption key K_c length is reduced, and a new key K'_c is derived. The polynomial modulo operation $K'_c(x) = g_2(x)[K_c(x) \bmod g_1(x)]$ is used, where $\deg(g_1(x)) = 8L$ and $\deg(g_2(x)) \leq 128{-}8L$. The $g_1(x)g_2(x)$ and $g_2(x)$ polynomials are specified by the Bluetooth system specifications.

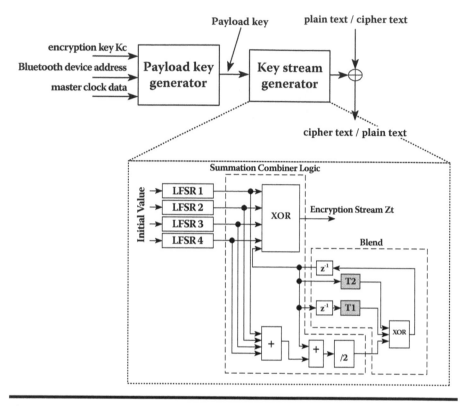

Figure 12.10 E_0 function architecture.

12.5 Bluetooth Enhanced Security Layer (ESL) Hardware Implementation

This section presents a new ESL for improving the security of Bluetooth technology. The security level is increased by replacing the original Bluetooth encryption with a design based on advanced encryption algorithm (AES) [8]. ESL adds MACs (media access controls) to the transmitted data for cryptographic integrity protection. The proposed layer is added on the top of the standard Bluetooth controller interface, which allows integrating it as an additional module into any standard Bluetooth chip or as a software layer into a host.

This section also presents a prototype implementation of ESL. For high performance and low energy, the security processing of ESL is implemented in hardware. The design supports the new Counter mode with Cipher block-chaining Message authentication code (CCM) encryption mode [9], which is adopted into several wireless IEEE standards, e.g., 802.11i. Thus, the same hardware design can be utilized in other standard implementations. In addition to the enhanced security, the

ESL implementation offers an Application Programming Interface (API) for a Bluetooth device by hiding the low-level Bluetooth commands from the application.

12.5.1 Enhanced Security Layer (ESL)

In order to overcome the shortcomings of the Bluetooth encryption algorithm and fix the lack of cryptographic integrity protection, ESL for Bluetooth has been designed. ESL replaces the proprietary encryption with AES and adds MACs to the transmitted data. It supports four well-scrutinized encryption modes with different characteristics and processing requirements. The application may choose the preferred one according to its security requirements. In addition to the increased security, ESL hides the security processing and low-level HCI operations and provides a high-level application programming interface (ESL API) for straightforward application development. To further improve the level of security, the design allows extending ESL to support, e.g., enhanced authentication and key exchange techniques. For the interoperability with devices that do not contain the enhanced features, the support for the standard Bluetooth security is also included in ESL.

The ESL architecture is presented in Figure 12.11. As shown, ESL is placed above HCI. Generally, Bluetooth technology is provided as fixed chips, which implement the Bluetooth functionalities below HCI. Application developers have only access to the Bluetooth controller through the standard HCI. Therefore, in order to improve the security, the most universally applicable method is to add the enhancements above HCI. This way ESL can be integrated as an additional module into any standard Bluetooth controller or host.

The same method is used, for example, in the implementations of the redesigned 802.11 Wireless Local Area Network (WLAN) authentication (802.11i). Instead of reimplementing WLAN cards, the old authentication is disabled and the existing cards are only utilized for transferring higher layer data. The new authentication is

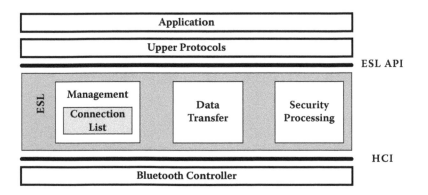

Figure 12.11   Architecture of Bluetooth ESL.

performed above the old WLAN link layer. This requires only reimplementing the software in the host computers.

Another advantage of placing ESL on the top of HCI and not into the baseband is that the method results in lower packet overhead. Added protocol fields are transmitted in the HCI data packets instead of every Bluetooth network packet.

12.5.2 Enhanced Security Design

The AES algorithm [8] utilized in ESL is a symmetric cipher that encrypts data in 128-bit blocks. It supports key sizes of 128, 192, and 256 bits. As several other block ciphers, AES consists of successive, similar iteration rounds. Depending on the chosen key size, the number of the rounds is 10, 12, or 14. Each round mixes the data with a round key, which is generated from the encryption key. Currently, the 128-bit key version is generally considered to provide adequate security and it is also utilized in this work. Decryption requires inverting the iterations resulting in at least partly separate data paths. However, the encryption modes of ESL only use the forward cipher.

Applying an encryption algorithm alone without a proper encryption mode is not secure. The counter mode (CTR) [10] is generally regarded as a good choice and it is also utilized in ESL. CTR has a proven security bound [14] and it provides most performance trade-offs for implementations. In CTR a block cipher produces a key stream from a secret key and a counter. The key stream is generated a block at a time by encrypting counter values until the stream length matches the data length. After each algorithm pass, the counter is incremented. The stream is exclusive-ored (XOR) with the plaintext to get the ciphertext and vice versa. If the data length is not a multiple of the block length, only the required bits of the last key stream block are used. It is important that the same counter value is used only once during the lifetime of a key. Another security requirement is that at maximum 2^{64} encryptions are to be performed per key.

The CTR mode requires that the encrypted data is accompanied with MAC. Otherwise, the bit manipulation attacks of section II-A still apply. In this work, MAC is computed using the cipher block chaining MAC (CBC-MAC) method. This allows utilizing the same algorithm for both encryption and data authentication. CBC is a feedback encryption mode in which the previous ciphertext block is XORed with the plaintext block before encryption. CBC-MAC operates in the same way, except that only the result of the last encryption is output as MAC. CBC-MAC is security proof [11].

ESL supports plain CBC-MAC (MAC mode) and two combinations of the CTR encryption and CBC-MAC computation. In the combined modes, MAC may be computed over the plaintext (MAC-then-encrypt mode) or over the ciphertext (encrypt-then-MAC mode). The phases utilize separate keys. In the MAC-then-encrypt mode, MAC is encrypted. The MAC mode is suited for applications in which privacy is not required. The combined modes provide both privacy and authenticity. In [12], it has been shown that the encrypt-then-MAC mode is generally secure. However, for example, in [13], it is suggested that it should be the plaintext that is authenticated. Adding the other mode to a design that implements one of the modes requires only little additional resources. Thus, both the modes are supported in ESL.

In addition, ESL supports a special MAC-then-encrypt mode, CCM [9]. Originally CCM was proposed to the IEEE 802.11i working group for improving the security of the IEEE 802.11 WLAN. It was also adopted. Later it had been sent for evaluation as a standard block encryption mode. The described components of the mode, CTR and CBC-MAC, have been well known for decades, but CCM is a new definition for their combined usage. The security of the CCM mode has been proven in [14].

Whereas the CTR encryption can process arbitrary length data, CBC-MAC requires that the input is padded to match a block boundary. In ESL, the last input data block is padded with zeroes if required in all the modes. The padding does not affect CTR. Instead, the en/decrypted output is truncated back to the original length. The MAC size can be chosen to be 64 or 128 bits, which allows trade-offs between the protection level and communication overhead. If the 64-bit MAC is chosen, the 64 least significant bits of the MAC output are discarded. MAC is appended to the HCI payload data.

In order to prevent repeated counter values, the CTR counter is composed of concatenated nonce and block counter values in all the combined modes. The nonce is a constant value for a transmission and the block counter is incremented for each data block within the transmission. The nonce is provided by the application. For example, [9] presents recommendations for choosing the nonce in CCM. Good practice is to include at least the sender's Bluetooth address and the transmissions sequence number in the nonce. In the presented work, the nonce size was chosen to be 96 bits to allow including sufficient amount of information in it. Thus, the block counter size is 32 bits. In addition to the nonce, in CCM there are fixed flags in the CTR input. In the presented approach, the flags are regarded as a part of the nonce.

Since ESL is located above the standard HCI, only the data placed in the HCI data packet payloads can be encrypted. However, the application can protect the known Bluetooth packet header fields (e.g., Bluetooth addresses) with MAC. In addition, the application must ensure that the nonce can be generated at the receiving device. The generation data can be predefined or transmitted in the HCI data packets. It does not have to be kept secret, as long as it is protected with MAC.

The format of an ESL packet, placed in the HCI data payload, is presented in Figure 12.12. The maximum size of the application data per ESL packet depends on the chosen MAC size. It is assumed that the complete nonce is transmitted in an ESL packet. If a combined mode is chosen, the ESL payload field is encrypted. The HCI payload is transmitted in a Bluetooth ACL network packet. If the HCI payload does not fit into a single network packet, the Bluetooth controller automatically

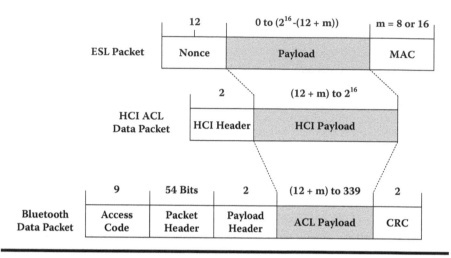

Figure 12.12 ESL, HCI ACL, and Bluetooth ACL packet formats. The field sizes without units are presented in bytes.

fragments the payload into several network packets. In Figure 12.12, the Bluetooth ACL packet is the ACL data packet with the largest payload size.

12.5.3 ESL Entities

As depicted in Figure 12.11, ESL consists of three entities: security processing, data transfer, and management. ESL is accessed through ESL API. The security processing entity performs the computations related to the security design of the previous section, i.e., the AES algorithm itself and the operations of the encryption modes. It appends MACs to the transmitted application data and verifies the received MACs. Failing packets are silently dropped. It also pads the input data to the block boundary and truncates the computed MACs into the desired length. The entity is not utilized if only the standard Bluetooth security is used.

The data transfer entity conveys application data between ESL API and the Bluetooth controller by constructing and decoding HCI data packets. If only the standard Bluetooth security is used, in transmission the entity places the application data into a HCI data packet payload field and forwards the HCI packet to the Bluetooth controller. Upon the reception of a HCI data packet, it decodes the application payload from the packet and gives it to ESL API. If the enhanced encryption is utilized, the payload to be transmitted in a HCI data packet is received from the security processing entity. Similarly, the payload of the received HCI data packet is first processed by the security processing entity.

The management entity controls the other entities and Bluetooth links. It initiates the Bluetooth controller, establishes and closes connections, and provides keys and other parameters (encryption mode and MAC length) to the security processing entity. The management entity constructs HCI commands and receives information from the Bluetooth controller in HCI events. The controller initiation consists of preparing the device to function as a slave or a master and, if the standard security is applied, providing the controller with the security parameters.

Another major responsibility of the management entity is to maintain a connection list, which contains handles to the established Bluetooth links and their ESL parameters, i.e., the encryption and MAC keys and chosen MAC length. If the enhanced security is not utilized for a link, the management entity sets ESL to bypass the enhanced security processing. Otherwise the entity advises the security processing entity to en/decrypt the HCI data payload and, if valid, give it to the data transfer entity for further processing.

The ESL processing and HCI are hidden behind the high-level ESL API. It provides procedures for device initiation, connection establishment, sending and receiving application data, and disconnecting. After the initialization, the security processing is transparent to the application. The interface is only requested to transmit or receive application data and provide nonces.

12.5.4 Prototype Architecture

The architecture of the ESL prototype is presented in Figure 12.13. The components implemented in the hardware (PLD) were Direct Memory Access (DMA),

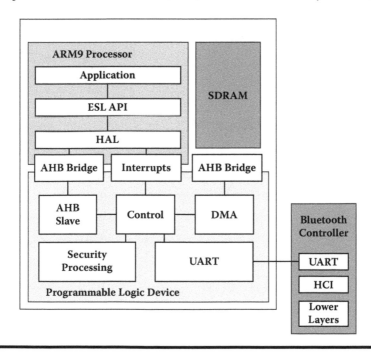

Figure 12.13   ESL prototype architecture.

UART (Universal Asynchronous Receiver/Transmitter) and its control, security processing and control, and processor-PLD communications. The security processing was implemented in hardware for high performance and low energy.

12.5.5 Security Processing Hardware

The security processing architecture is presented in Figure 12.14. Input data encoding and MAC value comparisons are performed by an external entity (i.e., ARM9). The internal signals are 128 bits wide, except the defined ones. The parameter updated internally is the block counter. The load signal sets up the module for reading a new encryption key, MAC key, and nonce values from the key_in and nonce ports. The keys are stored in Registers 6 and 7, nonce in Register 4. The nonce is

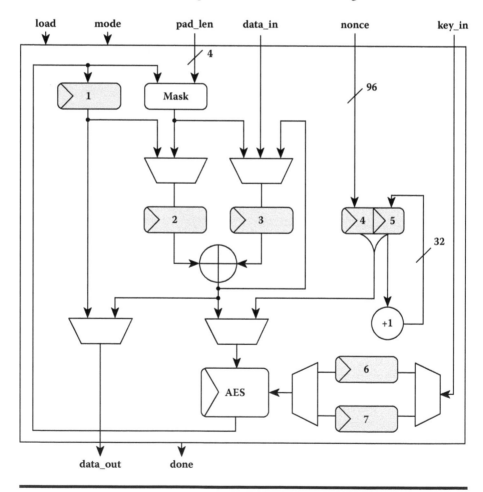

Figure 12.14 Security processing hardware architecture.

concatenated with the block counter in Register 5. The load port also allows maintaining the old key values and updating only the nonce. The block counter is reset by updating the encryption key or nonce.

Feeding a new data block is done through the data_in port. The block is stored in Register 3. The signal mode defines whether the module operates in the MAC, MAC-then-encrypt, encrypt-then-MAC, or CCM mode and sets the module to encrypt or decrypt. The pad_len signal is used for informing the number of padding bytes in the last input block. This information is required in the combined modes. The data_out port is used for outputting the en/decrypted data blocks as well as the MAC values.

In the MAC mode, MAC is computed using the MAC key. The chaining value in Register 1, obtained after processing the previous data block, is transferred to Register 2. Initially, the value is zero. After XORing the chaining value with the data block in Register 3, the result is processed by the AES core and the output is written back to Register 1. After the last data block, the contents of Register 1 are output and the register is reset. The MAC verifications are carried out in the same way.

In the MAC-then-encrypt mode, first, the design performs the MAC computation for a block with the MAC key as described. Then, the operation is switched to the CTR encryption with the encryption key. In CTR, the nonce and the block counter are fed to the AES core. Initially, the block counter is set to zero. The result is written through the mask block to Register 2 and XORed with the data block maintained in Register 3. Finally, the result is output and the block counter is incremented. After the last data block, the nonce and the block counter are processed once more and the result is written to Register 3 (through the mask entity). The contents of Register 1 are transferred to Register 2 and XORed with Register 3. The result is output as the encrypted MAC.

In the encrypt-then-MAC mode, the processing order of MAC-then-encrypt is inverted. The only difference is that after encrypting a data block, the output is also written back to Register 3 for the MAC computation.

In the MAC-then-encrypt decryption, the processing is mainly the same as in the encrypt-then-MAC encryption. After the last data block, the received, encrypted MAC is input through data_in to Register 3. The nonce and the block counter are input to the AES core. XORing the AES output with Register 3 yields the decrypted MAC, which is output for comparison with the earlier output, computed MAC. The encrypt-then-MAC decryption processing is the same as the MAC-then-encrypt encryption processing. The received MAC is not input since it is already in the plaintext form.

The CCM mode operation is similar to the MAC-then-encrypt mode. The encryption and MAC keys are set to the same value. The difference is that in the CCM mode the block counter starts initially from 1 and the MAC value is encrypted/decrypted with the block counter value zero.

If the data length is not a multiple of 16 bytes, the last output en/decrypted data block has to be truncated to the original length. This implies the need for the mask entity in the encrypt-then-MAC encryption, MAC-then-encrypt decryption, and

CCM decryption. Before XORing the last key stream block with the data block, the bytes of the key stream block corresponding to the extra bytes (zeroes) of the data block have to be masked to zero. This way the XOR result of the input block and the last key stream block, which is used as the input in the MAC computation, has the correct padding (zeroes). The mask logic is implemented with a ROM, containing 16 masking entries of size 16 bytes, and an AND gate.

12.5.6   Communication

As shown in Figure 12.13, the external Synchronous Dynamic Random Access Memory (SDRAM) is used for data transfers between the Advanced RISC Machine (ARM9) and the Programmable Logic Device (PLD) in the implementation. The external memory is larger than the fixed-size Dual Port Random Access Memory (DPRAM) and it can be switched, which makes the ESL implementation scalable for processing larger amounts of data. By sharing SDRAM, the processor does not have to transfer the data from the data memory in SDRAM to DPRAM for the PLD usage. Instead, the PLD can access SDRAM directly, which is faster and lets the ARM9 perform other tasks concurrently. The DMA entity was implemented for the purpose in the PLD. It accesses the memory via an Advanced High Performance Bus (AHB) bridge. An UART entity was implemented in the PLD for transferring data between the development board and the Bluetooth controller.

The data transferred through SDRAM consists of HCI commands and data to the Bluetooth controller, HCI events and data from the controller, and the data to/from the security processing entity. The nonce, encryption and MAC keys, and the UART initialization data are also conveyed through the memory. The nonce and keys are only transmitted to the security processing entity when the values are initialized or changed.

After writing a HCI command or data to SDRAM, the ARM9 utilizes the other AHB bridge for initiating operations in the PLD. The control entity receives the processor requests through the AHB slave in Figure 12.13. The slave contains logic for interfacing the AHB bus as well as control and memory address registers to which the ARM9 requests are written from the bus. Depending on the request, the control entity begins an UART transmission of a HCI command or the en/decryption of data. ARM9 is interrupted after an operation is finished. The processor reads the reason for the interrupt from the AHB slave.

When a HCI data packet is received from the Bluetooth controller, it is written to SDRAM by DMA and the ARM9 is interrupted. If the packet payload is not encrypted, the processor decodes the packet and gives the ESL payload to the application. Otherwise, it provides the security processing entity with the keys and nonce for decrypting the payload. When the payload is decrypted, the ARM9 is again interrupted. If a MAC scheme is used, the processor verifies that the received and computed MAC values match and conveys the data to the application.

Each HCI command has a corresponding event (acknowledgment) with which the Bluetooth controller replies to the command. In addition, the network operations trigger events. For simplicity and removing unnecessary memory accesses, the PLD control entity filters out the events uninteresting to the processor.

The communications between Bluetooth devices utilize the access control list (ACL) link. The used ACL packet types can be defined with a HCI command.

12.5.7 Software Interfaces

In Figure 12.13, hardware abstraction layer (HAL) implements the ESL functionalities on the ARM9 side. It constructs HCI commands and data and reads HCI events from SDRAM as well as decodes payload data from the HCI data packets. It also handles the interrupts initiated by the PLD. HAL controls the security processing entity and UART by modifying the memory mapped control and address registers in the AHB slave. HAL allows utilizing the Bluetooth's own security features as well as choosing among the enhanced security implementations. If an enhanced security mode is utilized, upon the reception of a HCI data packet, HAL compares the MAC values. It also pads the input data and truncates MACs.

The ESL API implementation provides procedures for initiation, connection management, and sending and receiving application data. A pseudo-code example of the API usage is presented in Figure 12.15.

In the example, first, the Bluetooth device is initialized. It is defined that the standard security features are not used. The InitBd procedure returns the unique Bluetooth address of the controller and the maximum size of the payload for a single transmission. After the initialization, the device is able to operate in the slave

```
void main() {
...
//initialize Bluetooth device:
//authentication disabled, no link key type defined, no PIN input
InitBd(FALSE, NULL, NULL, OwnBdAddress, payloadMaxSize);

//set device into master and scan devices nearby
PrepareMaster (numberOfResponses, deviceArray);

//set values for Bluetooth's own link key and for the keys of the end enhanced
//encryption and data authentication, choose encrypt-then-MAC
SetEncMode (NULL, encKey, macKey, ENC_THEN_MAC);

//create connection to the first found device with the security parameters above
ConnectToBd (deviceArray[0], connectionHandle);

//send data to the connected device
TransmitData (connectionHandle, payloadSize, payload, nonce);

//close connection
Disconnect (connectionHandle);
}
```

Figure 12.15   A pseudo-code example of using the ESL API implementation. Values for the parameters in italics are returned by the procedure calls.

mode. Next, the role of the device is switched to the master mode. The Prepare-Master procedure also scans for devices in the range and returns a list of found device addresses. Before connecting to a device, the parameters for the enhanced security features are set. Successful connection creation returns the handle of the created link. Application data can be sent over the link with a single procedure call. The data size must respect the maximum payload size defined in the initialization. Finally, the connection is closed.

Even though not utilized in the example, each procedure also returns a Boolean value that informs whether the performed operation was successful or not. For example, if the transmit buffer of the Bluetooth controller is full, the Transmit-Data procedure fails. Changing the security parameters requires closing the link and calling the InitBd and/or SetEncMode procedures again. The application must implement a separate callback procedure for receiving data. Upon the reception of a data packet, the procedure is automatically called by the ESL API implementation. Similarly to the standard Bluetooth, the implemented software enables up to seven simultaneously active Bluetooth connections.

12.6 Bluetooth Security Implementation Based on Software-Oriented Hardware–Software Partition

In this section, a hybrid implementation in software and hardware of the Bluetooth security algorithm is presented. Software-oriented and hardware–software partitioning methods are applied to this design and resource-sharing scheme is also adopted. The partitioning and resource sharing reduce hardware resources, though the timing requirement is fulfilled.

12.6.1 Constructing Block Ciphering Functions in Software

In the present approach, the block ciphering functions have been implemented in software. E_1, E_{21}, E_{22}, and E_3 functions are analyzed and broken down to subfunctions. Ar and A′r functions are computed from the same function with different mode values 0 and 1, respectively. They are basic functions for constructing the hash function. Figure 12.16 shows the relation of above functions.

E_x functions ($x = 1$, 21, 22, and 3) can be made with these basic subfunctions, Ar, A′r, and hash functions. E_1 and E_3 functions are hash functions themselves with L values, 6 and 12, respectively. Namely, E_1 (RAND, K, address) equals hash (RAND, K, address, 6), and E_3 (K, RAND, address, ACO) equals hash (K, RAND, COF, 12). E_{21} and E_{22} functions use A′r function whose input parameters are modified from original input parameters. Therefore, E_{21}(PIN, address) equals A′r(X(RAND), Y(address)), and E22(PIN, RAND) equals A′r(X′(PIN), Y′(RAND)).

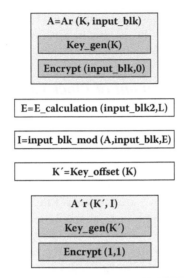

Figure 12.16 Constructing subfunctions. Ar, A'r, and hash functions.

12.6.2 *Timing Analysis of Block Ciphering Functions*

Regarding the system level analysis, the subfunctions were simulated on MS-Windows®-based Pentium® PC and measured each relative running time. The calculation time of key_gen function or t_key is 16.13 msec, and the calculation time of encrypt function or t_enc is 7.31 msec. Ar or A'r function needs t_Ar or 23.72 msec in its calculation, and the hash function needs t_hash or 47.44 msec. Because (t_key + t_enc) is about 98.82/100 of t_Ar and the double of t_Ar is 99.54/100 of t_hash, t_key and t_enc are critical calculation times in Ar and hash functions.

12.6.3 *Timing Constraint of Block Ciphering Functions*

Generation of each key and authentication are accomplished by link manager protocol (LMP) between Bluetooth devices. For example, if devices have a link key and want to authenticate each other, the verifier sends a random value with LMP_au_rand command in link manager level, and claimant returns SRES value, which is calculated with received random value using LMP_sres command. The verifier checks correspondence of its SRES and received SRES from claimant. LMP_sres is a response command of a LMP_au_rand command. The Bluetooth specifies the allowable response time of link manager command as 30 sec. Because within 1 slot time or 0.625 usec, the link manager command can be translated to the link control command and the link control command can operate the baseband block; the time constraint of each E_x function can be 29 sec to satisfy the Bluetooth specification. Even though the specification allows 30 sec to receive response command in link manager level, we set time constraint of each E_x function to 30 msec, 1,000 times smaller than what is given in the specification, in our design for fast key generation and authentication.

12.6.4 Timing Measurement of Block Ciphering Functions in Embedded MCU

In this Bluetooth design, it has used a Multipoint Control Unit (MCU) based on the SE3208 core, which is one of 32-bit extendable instruction set computer (EISC) processors made by Advanced Digital Chips, Inc., Korea [15,16]. EISC possesses advantages over existing Complex Instruction Set Computer (CISC) and Reduced Instruction Set Computer (RISC) architectures. EISC can represent any length of an operand without variable length instruction eliminating inefficiencies caused by difficulties in processing variable instruction decoding. EISC's fixed length instruction set maximizes cost/performance efficiency and at the same time offers flexibility and power through the extendable register and extension flag, which increase the code density while allowing a simple 16-bit-based instruction set. To measure the timing of the block ciphering functions in a 32-bit EISC processor, it uses Jupiter, which includes the same 32-bit EISC processor that it was intended to use [17]. The operating clock of Jupiter is set to 20 MHz, which is the same system clock of our Bluetooth design. The operating times of E_1, E_{21}, E_{22}, and E_3 functions are 22.462 msec, 11.842 msec, 11.516 msec, and 22.584 msec, respectively. Because these values satisfy the defined timing constraint of 30 msec, the block ciphering functions can be operated in EISC MCU without any replacement to hardware block.

12.6.5 Constructing and Analysis of Stream Ciphering Function

Bluetooth uses stream ciphering function, E_0 to encrypt or decrypt the payload of the packets. The E_0 function includes the payload key generation function to generate key stream. It means the E_0 function consists of the payload key generation function and key stream generation function. At the system level analysis, payload key generation function and the key stream generation function take 17.82 msec and 93.74 msec, respectively. In this analysis, because one operation of the key stream generation function makes 128 bits, each bit takes about 1/128 of 93.74 msec or 0.732 msec to generate.

Bluetooth specification requires that a payload key should be generated for each payload. If Bluetooth operates with DM1 packet type, it can transmit maximally 17 bytes or 136 bits in one slot timing, 0.625 msec. It needs one calculation of payload key generation and 136 bits generation of key stream. The payload key generation function needs 89.10 msec and the 128-bit key generation function takes 107.70 msec on an EISC processor with 20 MHz clock. Because neither the payload key generation nor key stream generation can be run within one slot time, 0.625 msec, all of them should be designed as hardware modules.

The payload key generation function has 128-bit modulo multiplication and 128-bit modulo division. Because each module takes 128 clocks in hardware module, the payload key generator function is operated with 256 clocks. Our Bluetooth design has a 20 MHz clock for EISC processor operation and a 1 MHz clock for the baseband operation. Even with the 1 MHz clock, the payload key generation can be finished in about 0.26 msec. The key stream generator needs 40 clocks for initialization of LFSRs and 1 clock for each stream bit. For the transmission of a 17 bytes-DM1 packet, the E_0 function in hardware takes 432 clocks, which consist of 256 clocks for payload key generation, 40 clocks for initialization of LFSRs, and 136 clocks for the payload of 17 bytes. With 1 MHz baseband clock, E_0 function is completed in 0.432 msec, which is less than one slot time.

12.6.6 Implementation Results

Bluetooth security modules are implemented in hardware and software as shown in Figure 12.17. E_0 function is designed in hardware modules, which are the payload key generator and the key stream generator. E_1, E_{21}, E_{22}, and E_3 functions are programmed in C language, compiled and downloaded to program ROM. It is verified on wireless SoC platform with Jupiter from Advanced Digital Chips and Virtex^{TM}$_2$ FPGA from Xilinx. As in the analysis of the previous section, Safer+ or Ar function consists more than 98/100 of each authentication or key-generation function.

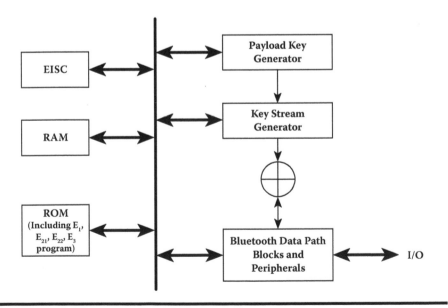

Figure 12.17 Block-diagram for implementation of Bluetooth security. E_0 is implemented in hardware and the other security functions are implemented in software.

By replacing only Safer+ function in hardware, operating time can be reduced considerably. With verilog code of Safer+ provided by Cylink Corporation, as shown in Figure 12.17, it has been constructed in hardware module with 1 MHz clock, which is the same as other Bluetooth baseband modules. C Program for Bluetooth Security is just changed from calling Safer+ subroutine or Ar in software to calling it in hardware. It executes E_1, E_{21}, E_{22}, and E_{23} functions in 0.887 msec, 0.337 msec, 0.287, and 0.851 msec, respectively.

12.7 Conclusion

In this chapter, an analytical description of the hardware architecture and implementation of the Bluetooth security, dictated by the official specifications was presented. Also, an implementation of a new enhanced security layer (ESL) is presented. The security level is increased by replacing the encryption with AES and adding integrity protection. The security processing is implemented in hardware for high performance. Finally, a partitioning hardware and software implementation was introduced.

ESL significantly improves the Bluetooth security by replacing the proprietary encryption with an AES-based design and adding cryptographic message integrity verification. In addition to the security enhancements, an easy-to-use API was designed. The API implementation offers an application developer-simple access to the wireless link and transparent security processing. The throughput of the implemented security processing hardware is 214 Mb/s with the modes offering the highest security level. This implies only a negligible processing latency, which is also lower than that of the standard security design.

Comparing the hardware implementation with the hybrid one (software and hardware), we notice that the proposed hardware modulo multiplier takes 256 clocks, whereas that in hybrid design takes 128 clocks. Even though overall hardware resources for E_0 function are similar, the number of slices and slice flip-flops in hybrid design are less than in the strict hardware implementation results, but the number of function generators in hybrid is larger than in hardware. The hardware resources of hybrid design are about 1/20 times than that of strict hardware. The number of slices, slice flip-flops, and function generators of hybrid and strict hardware are 770, 1,219, and 1,100 and 19,905, 2,600, and 42,764, respectively.

Security in pervasive computing is a complex issue and is getting a lot of negative reactions through poor implementations being brought to market (i.e., IEEE 802.11b). Many of the lower-level protocols are not secure and the use of the most secure higher-level protocols is limited. Bluetooth has the potential to enhance and extend the pervasive applications because it is well suited to pervasive devices and mobile applications.

Over the past three decades, we have labored under the constraint that secure cryptosystems required too much computation to be performed easily. These

constraints are disappearing. Moore's Law is producing general-purpose processors that can handle the necessary crypto functions in a negligible fraction of their
capacity. Tiny special-purpose chips can also be produced inexpensively for fulfilling the crypto demands of special applications. Thus, we are about to be freed from
the constraints of the past. This is even true for public key schemes. These algorithms, crucial for digital signatures and key management, do not require the communicating parties to possess a shared key that only they have. The computational
requirements of these methods are still considerably higher than for symmetric
ones, but progress in electronics is overcoming even this barrier.

Most of the single chip Bluetooth baseband implementations, which include a
low performance general-purpose processor, implement only the data encryption in
hardware. But, in time-critical applications where a fast connection solution is necessary and in devices with very high processing tasks, the hardware implementation
of the key generation mechanism and the authentication is also preferable.

References

[1] Bluetooth Special Interest Group, *Specification of the Bluetooth System*, vol. 1.1,
February 22, 2001. Available: http://grouper.ieee.org/groups/802/15/Bluetooth/
[2] J. Beutel, O. Kasten, A Minimal Bluetooth-Based Computing and Communication Platform, Technical Note, May 2001.
[3] R. Barber, Security in a Mobile World—is Bluetooth the Answer? *Computers and
Security*, 20, 5, Elsevier Advanced Technology, 2001, pp. 371–379.
[4] J.L. Massey and R. Rueppel, Linear ciphers and random sequence generators with
multiple clocks, *Advances in Cryptology: Proceedings of EUROCRYPT 84*, vol. 209
of Lecture Notes in Computer Science, pp. 74–87, Springer-Verlag, Heidelberg,
Germany, April 1984.
[5] P. Kitsos, N. Sklavos, K. Papadomanolakis, and O. Koufopavlou, Hardware implementation of the Bluetooth security, *IEEE Pervasive Computing, Mobile and Ubiquitous Systems*, 2, 1, Jan.–Mar. 2003, pp. 21–29.
[6] P. Hamalainen, N. Liu, R. Sterling, M. Hannikainen, T.D. Hamalainen, Design
and implementation of an enhanced security layer for Bluetooth, *The 8th International Conference on Telecommunications, ConTEL 2005*, Zagreb, Croatia, June
15–17, 2005.
[7] G. Lee and S.-Chong Park, Bluetooth security implementation based on software
oriented hardware–software partition, *2005 IEEE International Conference on
Communications, (ICC 2005)*, Seoul, Korea, May 2005.
[8] Advanced Encryption Standard (AES), Federal Information Processing Standards
Std. FIPS-197, 2001.
[9] D. Whiting, R. Housley, and N. Ferguson (January 2003) Counter with CBC-
MAC (CCM)—AES mode of operation. Available: http://csrc.nist.gov/Crypto
Toolkit/modes/proposedmodes/ccm/ccm.pdf.
[10] H. Lipmaa, P. Rogaway, and D. Wagner (October 2000) CTR-mode encryption.

[11] M. Bellare, J. Killian, and P. Rogaway, The security of the cipher block chaining message authentication code, *Journal of Computer and System Sciences*, 61, 3: 362–399, 2000.

[12] H. Krawczyk, The order of encryption and authentication for protecting communications (or: How secure is SSL?). In *Proc. 21st Annu. Int. Cryptology Conf. on Advances in Cryptology (CRYPTO 2001)*, Santa Barbara, CA, August 2001, pp. 310–331.

[13] N. Ferguson and B. Schneier, *Practical Cryptography*. John Wiley & Sons, New York, 2003.

[14] J. Jonsson, On the security of CTR + CBC-MAC. In *Proc. 9th Annu. Int. Workshop on Selected Areas in Cryptography (SAC 2002)*, St. John's, Newfoundland, Canada, August 2002, pp. 76–93.

[15] Adchips Inc., "EISC Architecture Smaller, Faster, Cheaper, all at Low Power," at http://www.adc.co.kr.

[16] H.-G. Kim, D.-Y. Jung, H.-S. Jung, Y.-M. Chio, J.-S. Han, B.-G. Min, and H.-C. Oh, AE32000B: A fully synthesizable 32-bit embedded microprocessor core, *ETRI Journal*, 25, 5, 337–344, October 2003.

[17] Adchips Inc., Jupiter Data Book, v 1.1, September 2002, at http://www.adc.co.kr.

Chapter 13

<hr>

Internet of Things: A Context-Awareness Perspective

Davy Preuveneers and Yolande Berbers

Contents

13.1   Introduction

The next wave in the era of computing will be outside the realm of the traditional desktop. In the Internet of Things* paradigm (IoT) [22], everything of value will be on the network in one form or another. Radio frequency IDentification (RFID) and sensor network technologies will give rise to this new standard, in which information and communication are invisibly embedded in the environment around us. Everyday objects, such as cars, coffee cups, refrigerators, bathtubs, and more advanced, loosely coupled, computational and information services will be in each others interaction range and will communicate with one another. Large amounts of data will circulate in order to create smart and proactive environments that will significantly enhance both the work and leisure experiences of people. Smart interacting objects that adapt to the current situation without any human involvement will become the next logical step to people already connected anytime and anywhere. With the growing presence of WiFi and 3G wireless Internet access, the evolution toward ubiquitous information and communication networks is already evident nowadays. However, for the Internet of Things vision to successfully emerge, the computing criterion will need to go beyond traditional mobile computing scenarios that use smartphones and portables, and evolve into connecting everyday existing objects and embedding intelligence into our environment. For technology to disappear from the consciousness of the user, the Internet of Things demands: (1) a shared understanding of the situation of its users and their appliances, (2) software architectures and pervasive communication networks to process and convey the contextual information to where it is relevant, and (3) the computational artifacts in the Internet of Things that aim for autonomous and smart behavior. With these three fundamental grounds in place, smart connectivity and context-aware computation via *anything*, *anywhere*, and *anytime* can be accomplished.

13.1.1   Software Architecture and the Internet of Things

Any software architecture, designed to be deployed within the setting of the Internet of Things, will need to foresee the increasing heterogeneity of devices and networks, and will have to provide for varying user and application requirements within diverse contexts of service provision. It will be essential for new architectures

---

* Internet of Things refers to a ubiquitous network society in which a lot of objects are "connected."

to integrate the latest technologies in the areas of software and hardware in order to assimilate as much as possible data and services available in the environment of the user. As service integration-related tasks will remain inevitable, it is no surprise that the industry is aiming for software solutions that make this integration a more tractable activity. The evolution toward integrated service orientation is already emerging in the ongoing convergence of Web services and telecommunication services [21]. Service-Oriented Architecture (SOA) [27] is an architectural style that enables the composition of applications by using loosely coupled and interoperable services. These often transaction-based services are described and exposed using open standards—in the case of Web services using WSDL, SOAP, and BPEL [40]—that are independent of the underlying programming language and platform. The telecommunications world has embraced the Service Delivery Platform (SDP) architectural approach to enable more flexible communication and collaboration services by evolving to all-IP (Internet Protocol) networks for the delivery of multimedia-enabled and location- or presence-aware services. Due to its event-driven communication paradigm, an SDP differs from the previous architectural style in that a SOA typically does not need to meet any real-time demands for its data-driven communication. Both SOA and SDP software architectures commonly provision service orchestration capabilities by exposing their services using open standards, and often share building blocks to handle noncore functionalities, such as authentication, identity management, and billing. As a result, converged services bring us the flexible interactions from the Web service paradigm, nicely integrated with the communication and collaboration services from the telecommunications world. Key to the success of the Internet of Things will be the ability to connect everyday objects, including RFID tags and other existing resources, to such service-oriented architectures so that it will lead to an overall flexible architecture that aims to create a pleasant user experience at the workplace, in public areas as well as in the home environment.

13.1.2 Context Awareness and the Internet of Things

Context awareness plays an important role in the aforementioned software architectures to enable services customization according to the current situation with minimal human intervention. Acquiring, analyzing, and interpreting relevant context information [11] regarding the user will be a key ingredient to create a whole new range of smart entertainment and business applications that are more supportive to the user. Although context-aware systems have been in the research epicenter for more than a decade [32,33], the ability to convey and select the most appropriate information to achieve nonintrusive behavior on multiuser-converged service platforms in mobile and heterogeneous environments remains a significant management challenge. Interoperability at the scale of the Internet of Things should go beyond syntactical interfaces and requires the sharing of common semantics across all software architectures. It also demands a seamless integration of existing computational artifacts (hardware and software) and communication infrastructures.

Only then can context information be successfully shared between highly adaptive services across heterogeneous devices on large-scale networks that consider this information relevant for their purposes.

13.1.3 Convergence as a Key Enabler for the Internet of Things

In summary, the Internet of Things is all about convergence, from connected computing using RFID and sensor technology to digital content and context-aware services, an observation that was also made in the digital lifestyle chapter of the *digital.life ITU Internet Report 2006* [23]. The success of the Internet of Things will not so much depend on the development of new technologies, but more so on connecting and integrating existing resources, ranging from small-scale objects, such as RFID tags, up to large-scale software systems that serve thousands of clients at a time. The goal is to create a software architecture that enables objects to exchange information through the Internet to achieve nonintrusive behavior and customized services. Part of the architecture's responsibility is to make sure that relevant information arrives at the right place in a way that the recipient understands what it receives. Here, we investigate how such an architecture can be designed by integrating innovative scientific results from related research domains with industrial technologies and practices with a proven track record.

In section 13.2, we describe the current state-of-the-art in context-aware computing and discuss the goals, motivation, and requirements that define the actual need for context awareness in order to achieve a better understanding of the concepts involved. We review the trends for its use in converging service-oriented architectures in section 13.3. To illustrate the benefits of introducing context awareness as one of the underpinning fundamentals of the Internet of Things, we also elaborate on our recent research activities at the architectural level in section 13.4. It illustrates how to integrate existing technologies and computational artifacts supporting context awareness as an enabling service amongst other building blocks in an overall service-oriented architecture for use in home appliances and online consumer services. By not reinventing the wheel, but instead leveraging on solid foundations from the software and knowledge engineering domains, nothing stands in the way for applications targeting the Internet of Things model to become a success.

13.2 State-of-the-Art on Context-Aware Computing for Nonintrusive Behavior

The notion of context is widely understood in the pervasive and ubiquitous computing domain as relevant information referring to the situation and circumstances in which a computational artifact is embedded. As such, context awareness is the ability to detect and respond to contextual changes. The goal of context-aware computing is to gather and utilize information to positively affect the provisioning of services that are considered appropriate for a particular person or device. Therefore, context information can only be considered useful if it can be interpreted. As context is a rather vague concept, we first mention how context has been defined by leading experts in the field before continuing to describe how context can be modeled, acquired, and used to achieve autonomous and nonintrusive behavior in a service-oriented architecture.

13.2.1 A Definition of Context

Many authors initially defined context information by enumerating types of information related to the user or application environment that seemed relevant. The term context was first used by Schilit and Theimer [32] to refer to "location, identities of nearby people and objects, and changes to these objects." Brown, Bovey, and Chen [6] have defined context as "location, identities of the people around the user, the time of day, season, temperature, etc." Ryan, Pascoe, and Morse [31] referred to context as "the user's location, environment, identity, and time." Dey [10] listed "the user's emotional state, focus of attention, location and orientation, date and time, objects and people in the user's environment" as elements being part of the definition of context. As the use of enumerations to describe context was too limited to analyze whether certain information could be classified as context, Dey et al. [1] provided the following more general and widely accepted definition that encompasses the previous ones:

> Context is any information that can be used to characterize the situation of entities (i.e., whether a person, place or object) that are considered relevant to the interaction between a user and an application, including the user and the application themselves. Context is typically the location, identity and state of people, groups and computational and physical objects.

Therefore, context involves relevant information on real world entities. This information needs to be described in a structured and easily extensible model to facilitate the sharing of collected information. Therefore, it is impossible to limit context information to a fixed set of attributes or properties.

13.2.2 Using RFID to Sense Context

RFID is an emerging technology for embedding sensing capabilities in everyday objects [39] and is gaining momentum as a popular means for automatic

96-BIT ELECTRONIC PRODUCT CODE (EPC)			
Header	EPC Manager	Object Class	Serial Number
01 8 bits	**0000A89** 28 bits	**00016F** 24 bits	**000169DCO** 36 bits

Figure 13.1   A passive RFID tag having 96 bits of memory to represent an EPC number.

identification and tracking in supply chain management. Each tag contains a unique ID number that can be read by an RFID reader. Active RFID tags have their own power supply to transmit their signal, while passive ones use the electrical current induced in the antenna during reception of the incoming radio frequency signal emitted by the RFID reader. The presence of an internal power supply helps to extend the range of operation and the amount of information that can be transmitted. The majority of passive tags typically have anywhere from 64 or 96 bits (to represent an electronic product code (EPC) [13] as shown in Figure 13.1) to 1 kb of nonvolatile EEPROM memory, while active tags have battery-backed memories as high as 128 kb and more.

The ability to store and remotely recognize tags at a high pace (in the order of hundreds per second) makes RFID a promising technology for identification and locating purposes in context-aware and pervasive computing [36]. RFID tags are mainly used for asset tracking and in inventory systems at libraries and shopping malls where they replace the older barcode technology. However, RFID technology is also being applied to the tagging of humans to identify them as well as locate their whereabouts. RFID enabled E-passports are issued by many countries, while implantable RFID chips are used to track patients in a hospital and access their medical records. Philipose et al. [28] and Smith et al. [35] also illustrate how RFID technology can be used to infer human activity. As such, it is clear that RFID provides added value to the domain of context-aware computing and to the Internet of Things paradigm in general for sensing identity and location.

13.2.3 Requirements for Representing and Exchanging Context

For humans, it is natural to communicate and interpret context. The goal of using open context specification languages is to simplify the capturing, transmission, and interchangeability of this context between systems. Different modeling approaches for context have been investigated in the past. A model consisting of key-value pairs is the most simple approach to represent context, but lacks the ability to structure information and only supports exact matching. Markup scheme models typically introduce a fixed structure and allow to express more complex relations, such as associations. These types of models are frequently used to capture user profiles (e.g., Friend of a Friend (FOAF) [5]) and device profiles (e.g., Composite Capabilities/Preferences Profile (CC/PP) [24] and User Agent Profile (UAProf) [16]). So far, ontologies, providing a specification of a conceptualization [17], seem to be the most promising way to go when modeling context [37,7,18,38]. Ontologies are well known in the knowledge representation community to model concepts and the relationships that hold among them and their semantic interpretation is universally accepted. Figure 13.2 illustrates a graphical representation of the

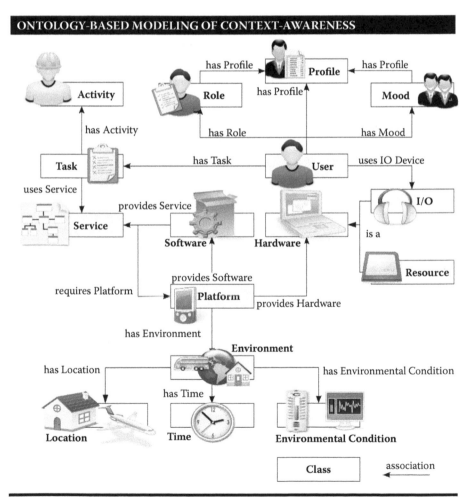

Figure 13.2   The CoDAMoS context ontology that models the user, platform, service, and environment concepts and interrelationships with other relevant classes.

Context-Driven Adaptation of Mobile Services (CoDAMoS) context ontology [30]. Many dedicated and XML-based languages exist to model ontologies, but the best known language is the Web Ontology Language (OWL) [20], a markup language for publishing and sharing data using ontologies on the World Wide Web. Following is a review of some of the main requirements for an open context specification to be used for exchange between computational artifacts in the Internet of Things:

- **Comprehensive domain coverage and terminology:** An open context specification language should make available a terminology that provides appropriate coverage and a comprehensive representation of a domain in order to model most of the concepts and terms needed for describing entities in a particular domain. Concepts may be modeled with multiple synonymous representations and may have hierarchical relationships to other concepts.

- **Semantic nonambiguity and expressiveness:** Information semantics involve the uniform interpretation of a concept. This requires a strict support for nonambiguity in the specification language to ensure that each concept in the terminology has only one meaning. Semantic expressiveness is the ability to easily enhance the knowledge domain using the semantic primitives of the specification language. More advanced semantic specification languages for context provide support to model specialization and inheritance relationships, aggregation, dependencies and constraints, etc.

- **Processing complexity:** Some languages require quite complex processing steps for both the data acquisition and processing (e.g., reasoning and inference) parts. For example, context models using Resource Description Framework (RDF) [4] or OWL [20] may refer to other remotely available context models to detail the semantics and relationships of concepts being reused. This, of course, is a good thing as it improves knowledge sharing, interoperability, and universal interpretation. However, it also increases the complexity to process context models. As the Internet of Things targets computational artifacts of different shapes and sizes, an appropriate balance should be found between processing complexity and language expressiveness.

- **Specification language interoperability:** There is no one single specification that will solve all issues with respect to context provisioning in an open fashion. For some purposes, the use of a particular domain-specific language may already be so pervasive that this in itself would be one of the reasons not to reimplement the same concepts in another more generic specification language. Therefore, the ease of reusing and integrating different dedicated specification languages is also an important concern.

For information to be processed automatically, it goes without saying that the markup needs to conform to a standard format that is accepted by as many parties as possible. Open standards are a hot topic at present for exchanging documents in a format that can be read and modified by different parties. A lot of attention is paid to XML formats as the Holy Grail for interchangeability. However, for context specification and exchange, it is not only important to be able to process the context information, but also to interpret the information. A large difference with the XML-based documentation formats, such as Open Document Format (ODF) [26] and OpenXML [12], is that the semantics of context information should be shared by all the parties involved, preferably within the context model itself. To this extend, XML only provides a tree-based structure to specify information. The interpretation is totally left to the user or the program.

Many open specification languages are general purpose description languages (e.g. RDF [4] and OWL [20]), while others target a very specific domain. For example, dedicated languages and specification formalisms have been proposed to model the hardware and software characteristics of mobile handheld devices (e.g., CC/PP [24] and UAProf [16]). Regarding RFID technology, it was only recently (2007) that the Electronic Product Code Information Services (EPCIS) standard [14] was ratified, effectively providing an XML schema binding and a binding to Simple Object Access Protocal (SOAP) over HTTP via a WSDL (Web Service Description Language) to the EPC information on an RFID tag. Some of these domain specific formats, such as UAProf, build upon a general purpose language like RDF, as illustrated in Figure 13.3. Not only does the expressiveness of

NOKIA N95-1 USER AGENT PROFILE

```
<rdf:RDF xmlns:rdf="http://www.w3.org/1999/02/22-rdf-syntax-ns#"
  ...
  <rdf:Description rdf:ID="Profile">
    <prf:component>
    <rdf:Description rdf:ID="HardwarePlatform">
      <prf:Vendor>Nokia</prf:Vendor>
      <prf:Model>N95-1</prf:Model>
      <prf:BitsPerPixel>18</prf:BitsPerPixel>
      <prf:ColorCapable>Yes</prf:ColorCapable>
      <prf:CPU>ARM</prf:CPU>
      <prf:ImageCapable>Yes</prf:ImageCapble>
      <prf:Keyboard>PhoneKeyPad</prf.Keyboard>
      <prf:NumberOfSoftKeys>2</prf:NumberOfSoftKeys>
      <prf:ScreenSize>240×320</prf:ScreenSize>
      <prf:ScreenSizeChar>15×6</prf:ScreenSizeChar>
      <prf:SoundOutputCapable>Yes</prf:SoundOutputCapable>
      <prf:TextInputCapable>Yes</prf:TextInputCapable>
      <prf:VoiceInputCapable>Yes</prf:VoiceInputCapable>
    </rdf:Description>
    </prf:component>

</rdf:RDF>
```

Figure 13.3   An partial description of a wireless device using the RDF-based User Agent Profile specification.

these languages differ, they also distinguish themselves in managing or processing complexity. As such, there is no single one-fits-all context specification language.

13.3 Trends in Context-Aware Computing within Service Orientation

Though efforts in context awareness research only seem to reach the public at a slow pace, we notice that the need for knowledge engineering methodologies becomes a growing concern in the design of software architectures when building user-aware services. These evolutions become apparent in both the SOA world of IT services, as well as for the SDPs of the telecommunications domain. Any software architecture for the Internet of Things will face the same concerns, but at a much larger scale as the number of information-sharing entities will, in the end, outnumber human communication.

13.3.1 Context-Enabled Service Oriented Architecture and the Semantic Web

Service Oriented Architecture (SOA) represents the current state-of-the-art in software architecture for the rapid deployment of new services. It enables the creation of new services or applications by connecting together existing services and proposes functions to manage the service life cycle. The main motivation behind SOA for a company is to create a business-aligned architecture to better react to changing customers' needs. If the market changes, new services can be created by reusing existing services and by developing new services where needed. A SOA also enables a company to leverage previous infrastructure investments, by exposing legacy services as traditional services in the architecture: A SOA increases code reuse and modularity.

A service-oriented architecture targeting the growing proliferation of RFID-equipped devices within the Internet of Things, where each computational artifact may access multiple services within diverse contexts, will become a daunting task without proper identity management. The use of context awareness helps to provide more fine-grained access control to information and services. One illustration of this trend is presented in the work of Ardagna et al. [3] where position, movement, and interaction predicates are introduced in authorization policies relying on Groupe Spécial Mobile (GSM)/third-generation (3G) technologies to achieve location-constrained access. With the introduction of the E-passport, fitting in RFID information into policy enforcement rules is just a small step. For managing the information flow in the Internet of Things, context awareness is becoming a much larger concern in SOA than it is at the moment.

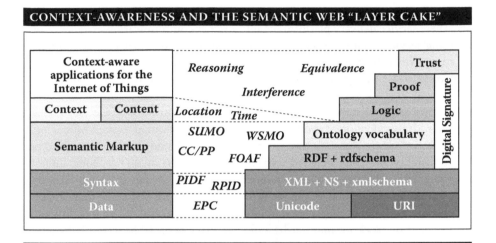

Figure 13.4 Aligning context awareness and standardized specifications for the Internet of Things along the Semantic Web "layer cake" (Courtesy of T. Berners-Lee).

Another emerging trend is coming from the Semantic Web community. Although the WSDL [8] has been very successful as a syntactic specification language to interact with Web services using different backend environments, it does not provide any support for a consistent interpretation of the Web service content. The use of ontologies has already gained much attention for the semantic modeling of context, as illustrated in Figure 13.4. The application of ontologies is also finding its way in new Web service specifications, including OWL-S [25], METEOR-S [34], WSDL-S [2], WSMO [9,15], and others, to augment service descriptions in order to automate their discovery, composition, and invocation in pervasive computing environments. The services, that were previously only described by their public invocation interfaces, are now enriched with machine-interpretable semantics by referring to and reasoning on established ontologies that encapsulate the meaning of the service. Also, for the Internet of Things, the use of ontologies within the architecture may help to make sense of it all, for both the context information as well as the services being offered.

13.3.2 Context-Enabled Service Delivery Platforms

During the last decade, the telecommunications world has transformed radically due to the shift toward Internet-style infrastructures for the delivery of multimedia-enabled communication services. Today's very popular, IP-based VoIP (Voice over Internet Protocol) systems are the result of the evolution from monolithic and isolated infrastructures to service delivery platforms that adopt the IP network as their communication network and base their operations on open standards and protocols. These service delivery platforms enable a richer communication experience by providing services that go beyond leaving a voice message when your friend does not answer his cell phone. With conventional communication, you have no idea when your friend will get your message, perhaps leaving you no choice but to call back several times. If you only knew where your friend was at that moment....

Presence and availability are just two of the main features that these SDPs offer, together with the adoption of another popular Internet application: Instant Messaging. A presence service enables a selective group of users, the "buddy list," to be notified of your availability and your preferred way of communication (mail, telephone, instant messaging, etc.). This information is passed on using the Session Initiation Protocol (SIP). Similar to HTTP and SOAP for Web services, the SIP [19] is the most important protocol for Internet telephony, supporting user, session, and service mobility, and a major building block for multimedia and voice services on IP Multimedia Systems (IMS). SIP is under continuous expansion to support new types of communication and information that were beyond its scope when it was initially defined. Many of these add-on SIP standards are used to sketch the context of the users and their appliances, or to notify remote subscribers of a change in this context:

■ **SIP Specific Event Notification:** This is an extension that allows SIP nodes to request information from remote nodes. Being a generic package, it can be used to notify and convey any contextual change to whom it may concern when such an event has occurred.

■ **Presence Information Data Format (PIDF):** This is an XML document that provides presence information about a presentity, including status information and optionally contact addresses, timestamps, and textual notes. An example is given in Figure 13.5.

PRESENCE INFORMATION DATA FORMAT

```
<?xml version="1.0" encoding="UTF-8"?>
<presence xmlns="urn:ietf:params:xml:nspidf"
  entity="pres:john@doe.com">
 <tuple id="123456">
  <status>
     <basic>open</basic>
  </status>
  <contact priority="0.75">tel:+32987654321</contact>
  <timestamp>2007-03-15T17:26:21Z</timestamp>
  <note>I am busy right now ...</note>
 </tuple>
</presence>
```

Figure 13.5   Specifying presence within SIP using the basic PIDF format.

- **Rich Presence Information Data Format (RPID):** This format builds upon PIDF and adds elements that provide additional information about the presentity* (location, mood, activity, etc.) and its contacts.
- **Other related SIP standards:** Common Profile for Preference, Common Profile for Instant Messaging, Contact Information in Presence Information Data Format, Timed Presence Extensions to the PIDF, etc.

We can observe that, given the list of SIP specifications and draft recommendations, presence is an important concern in telecommunications. In supply chain management, where RFID technology typically has a strong foothold, presence and availability play an important role as well. Although products will not directly communicate their presence to all interested parties by themselves, it is clear that the flexibility of the SIP signaling protocol makes it an excellent candidate to deliver real-time and up-to-date context information to all subscribers.

13.4 Service Provision in a Context-Aware, Converged Service Architecture

Interoperability between Web service-and telecommunication-oriented architectures is a complex issue. A service in SOA is essentially transaction-based and relies on an architecture that often builds upon the functionality of an enterprise service bus (ESB) to provide message brokering, routing, data translation, and transformation. A service within a SDP needs to deal with a multitude of point-to-point connections and short-lived events that must be generated, propagated, and processed with a low latency to guarantee a minimal quality of service level. The proposed architecture for the Internet of Things, therefore, integrates both kinds of services at the session level because that is where the interaction between two services starts and where it can benefit the most from context information. As Web services and telecommunication services use similar though competing standards for information exchange; an enterprise service bus is used inside the architecture to manage the message conversion and routing during a session. A high-level conceptual overview of the architecture is given in Figure 13.6.

The goal of introducing context awareness in a converged service-oriented architecture is to simplify the discovery of information about an entity when this information cannot be easily searched, is not made explicit in the required format, or needs aggregation with other information sources before it can be used. Our context management system [29] represents itself as an enabling service in the converged service-oriented architecture, in a similar way as billing and policy enforcement do. Neither of these services provide any functionality directly to the client, but do so to other enabling and end-user services. Similar to billing and policy

---

* Presentity: A presentity (presence entity) provides presence information to a presence service (RFC2778).

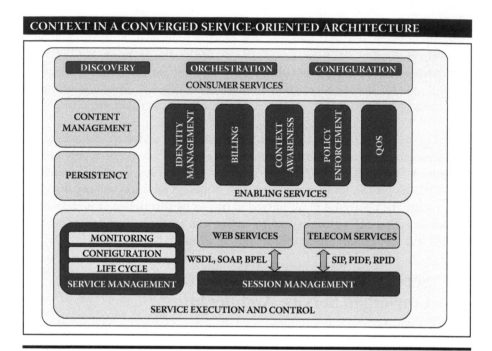

Figure 13.6   Context as an enabling service in a converged service-oriented architecture.

enforcement, special care is taken to make sure that context information belonging to different entities is processed and stored separately. From an architectural point of view, however, our context-awareness enabling service acts as any other service whose life cycle in the system needs to be managed.

13.4.1   Context Management as an Enabling Service

Context management from an enabling service perspective involves the following functional requirements, no matter whether the context information is used locally or needs to be delivered to a remote computational artifact somewhere on the network.

- **Context Acquisition:** This function gathers information from RFID tags, sensors, user profiles, presence servers, or other information providers in the system itself or on the network. In general, it monitors for context that is changing and collects all information that remote entities push to the system using SOAP, SIP, or another domain-specific communication protocol.
- **Context Storage:** A context repository ensures persistence of context information. It saves relevant information in a way that queries and information updates can be handled efficiently without losing the semantics of the data. It, therefore, preserves the information itself, e.g., 'Age=23;' to which entity it belongs, e.g., a person or a device; when and by whom it was acquired; and

how the information relates to other instantiated context concepts by referring to the right ontology in the semantic knowledge base, also stored in the context repository.

■ **Context Manipulation:** This part aggregates and reasons on context information to provide information in a suitable format when needed. Besides a reasoning engine that exploits semantic relationships within ontologies, the enabling service also provides adapters that transform context information to change the way information is represented. For example, a temperature expressed in °C can be translated into °F using straightforward conversion rules. Classification sacrifices accuracy of low-level information for the sake of obtaining more meaningful information. For example, a geographical location of longitude and latitude coordinates can be converted into a city representation. As many application domains propose similar though competing standards, part of the manipulation is also to bridge the standards interoperability gap where possible.

A high-level overview of the building blocks of the context-awareness enabling service is given Figure 13.7.

13.4.2 Conveying Context and Distributed Storage

After having consolidated context information from various acquisition systems (sensors, people, devices, etc.) and remote input channels (services, databases, etc.), the collected information may be selected for distribution to remote subscribers (people, services, devices). When information is gathered from or needs to be

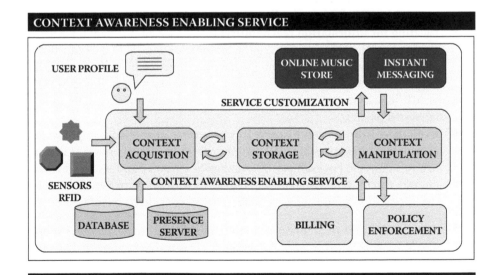

Figure 13.7 Building blocks of the context awareness-enabling service.

delivered to a remote entity, choices need to be made on where context will be stored. The obvious choice would be close to the area where it is considered relevant, and perhaps be duplicated if necessary. If context is stored locally on the device that sensed the information, it is readily available and round trips on the network to retrieve the information are avoided. However, if the information is also useful for other parties, then it makes sense to store the information remotely on a repository that is more easily accessible. The following concerns are taken into account whether information should be stored locally or remotely:

- **Scope of relevance:** Certain information can be relevant only for computational artifacts that interact in particular with local services. An obvious example is the intelligent home environment where the information that photovoltaic light sensors in the house provide, is not of interest outside the home environment.

- **Information sensitivity:** Tracking information retrieved from RFID tags could be of interest to multiple parties on the network and be made available on a remote server for further analysis. Personal information, on the other hand, is kept private on purpose and only shared with authorized entities. Although privacy issues are an important concern in context-aware computing, we will not go into detail here as privacy is already covered to a large extent in other chapters and in the specialized literature.

- **Information currentness and period of relevance:** Static and profiled information is likely to remain valid for longer periods. For example, the position of fixed beacons is information that is highly suited for distribution, as synchronization does not need to occur often. Volatile information on the other hand, such as the current system load, may no longer be accurate or relevant when it is finally remotely stored in a repository.

- **Caching for information reachability:** If some sensor information can be obtained quite easily, then storing the information may not be worthwhile. However, as some appliances might have no direct connection to context sensors that provide useful information, a repository with standard SOAP or SIP interfaces can collect and make the information accessible to all interested parties.

- **Exploiting information history:** In some cases, an appliance might want to exploit the historic values of a certain context attribute in order to derive new information. For example, by tracking the current position, the traveled distance can be derived and used to reimburse traveling expenses. Although the information is no longer accurate, it is still considered useful.

It is clear that the software architecture itself cannot always decide if context information should be distributed or not. The reason for this is that context is all about relevant and useful information and the software architecture must be informed by its clients about what information they consider relevant for their purposes.

13.5 Context Awareness from RFID to the End-User in the Internet of Things

The digital lifestyle chapter of the *digital.life ITU Internet Report 2006* [23] describes several futuristic scenarios toward context-aware services. We will briefly summarize them and then discuss how our converged service-oriented architecture would accomplish the tasks in these scenarios if the architecture would be fully implemented and deployed on a large scale.

1. You are walking around in the city and your favorite shop calls you on your mobile phone. It has detected that you are in the neighborhood and wants to promote some goods you might be interested in.

2. You find the shirt of your dreams, but it is not available anymore in your size. You scan the RFID tag to look for branches that have the shirt in stock and request driving directions to the nearest branch.

3. After your shopping expedition, you head back to your car and while going home you run low on fuel. The GPS navigation system in your car informs you about nearby gas stations.

4. The next morning, your umbrella notifies you as you leave for work, that the weather forecast appears to predict rain for this evening. So, you better be prepared and take along that umbrella.

5. At work, your telephone handset detects a high body temperature. It seems you are a bit feverish. The display shows a warning, along with healthcare information and contact information of a nearby doctor.

Most of the above scenarios rely on location awareness of the user. This location could be obtained using RFID or Bluetooth technology installed in the cell phone of the client, or from a GPS navigation system in the car. The types of location information is different (identifier, MAC address, latitude, and longitude) and will likely need to be converted first by the context-awareness-enabling service before it can be used by one of the end-user services. Presence awareness comes into play in scenario 1 and 5. The context-awareness-enabling service verifies first using the SIP protocol if the person in question is available and what kind of communication he currently prefers, and then sends a message to the corresponding address. User profiles are used by the shop owner to link an identity with previously purchased goods, and at client side to store the home and work address, the size of your shirt, SIP contact information, etc. Some context information conversion is needed to convert the current location to a four-letter METAR weather station identifier to get the local weather report at home and at work. Sensor technology is used to check the amount of fuel, to check the body temperature, to detect movements at the front door of the house, etc. The second scenario could be an orchestration of Web services that are semantically described using OWL-S. After using SOAP to send the EPC of the shirt and the size you need to the Web service, you obtain a list of addresses. For each of these addresses, another online route planning service is used to compute the distance to the nearest branch.

The above scenarios provide a mixture of Web service and telecommunication service orchestrations, context awareness, RFID and sensor technology, local and remote services, SIP and SOAP communication, and so on. Although a real-life deployment of the architecture has not been done to test the feasibility, we are confident that our current architecture can accomplish each of these tasks. All the requirements have been covered in the previous sections and interoperability support is built into the architecture to make these scenarios work.

13.6 Conclusions

The Internet of Things is all about convergence and integration of the latest advancements in the research areas of software and hardware with industrial technologies invented many decades ago. Indeed, although RFID is one of the cornerstones of this new computer paradigm and becoming more widespread nowadays due to its low production costs, its invention stems from the World War II period when it was developed for military purposes. In this chapter, we have highlighted two concerns of great importance for the Internet of Things. One comes from the software engineering domain, namely the use of a good software architecture style for the design of any software system, and another one is related to the knowledge engineering domain and deals with context awareness, also one of the goals of the Internet of Things. Research carried out in the context-aware computing domain aims to create a pleasant user experience at the workplace, in public areas as well as in the home environment by simplifying human interactions with everyday services and making them less intrusive. We have discussed how RFID technology can contribute to context awareness as a way to identify and locate everyday objects as well as humans, for that matter.

We have argued why service orientation is a good architectural style for the Internet of Things because it leverages existing applications by exposing them as traditional services in the architecture. It increases reuse and modularity, speeding up the integration effort. We have proposed a service architecture that aims to bridge context awareness with the worlds of Web services and telecommunication services. Both application domains have contributed in their own way to the development of context-aware services by standardizing description formats and protocols to convey relevant information to whom it may concern. Standardization is good, only if not for the fact that many of them are similar and competing with one another. For this reason, we have introduced context awareness as an enabling service in the architecture, leveraging information from both areas at a central place. The context-awareness-enabling services is responsible for the acquisition, persistency, and manipulation of information to make sure that the right information can be delivered at the right place at the right time.

13.7 Open Issues

Despite the great potential of context awareness for the Internet of Things, there are still some hurdles to cross that may slow down the adoption of RFID and other emerging technologies. One of the repercussions of context-aware computing is the invasion of one's privacy. While the aim is to increase convenience by reducing the amount of interactions between users and devices, it is of utmost importance to not lose sight of providing a level of user control in order to set the minds at ease. Only when people are part of the Internet of Things and feel in control will this new way of computing have a chance of being widely adopted.

References

[1] G.D. Abowd, A.K. Dey, P.J. Brown, N. Davies, M. Smith, and P. Steggles. Towards a better understanding of context and context-awareness. In *HUC '99: Proceedings of the 1st International Symposium on Handheld and Ubiquitous Computing*, 304–307, London, 1999. Springer-Verlag, Heidelberg, Germany.

[2] R. Akkiraju, J. Farrell, J. Miller, M. Nagarajan, M.T. Schmidt, A. Sheth, and K. Verma. Web Service Semantics—WSDL-S, v. 1.0, November 2005.

[3] C.A. Ardagna, M. Cremonini, E. Damiani, S. De Capitani di Vimercati, and P. Samarati. Supporting location-based conditions in access control policies. In *ASIACCS '06: Proceedings of the 2006 ACM Symposium on Information, Computer and Communications Security*, pp. 212–222, New York, 2006. ACM Press.

[4] D. Beckett. RDF/XML Syntax Specification (Revised), February 2004.

[5] D. Brickley and L. Miller. FOAF Vocabulary Specification, 2005.

[6] P.J. Brown, J.D. Bovey, and Xian Chen. Context-aware applications: From the laboratory to the marketplace. *Pers. Comm., IEEE* (see also *IEEE Wireless Comm.*), 4(5): 58–64, 1997.

[7] H. Chen, T. Finin, and A. Joshi. An ontology for context-aware pervasive computing environments. *Knowl. Eng. Rev.*, 18(3): 197–207, September 2003.

[8] E. Christensen, F. Curbera, G. Meredith, and S. Weerawarana. Web Services Description Language (WSDL) v. 1.1, March 2001.

[9] J. de Bruijn, C. Bussler, J. Domingue, D. Fensel, M. Hepp, U. Keller, M. Kifer, B. König-Ries, J. Kopecky, R. Lara, H. Lausen, E. Oren, A. Polleres, D. Roman, J. Scicluna, and M. Stollberg. Web Service Modeling Ontology (WSMO), June 2005.

[10] A.K. Dey. Context-aware computing: The CyberDesk project. In *Proceedings of AAAI '98 Spring Symposium, Technical Report SS-98-02*, pp. 51–54, Menlo Park, CA, March 1998.

[11] A.K. Dey. Understanding and Using Context. *Personal Ubiquitous Comput.*, 5(1): 4–7, February 2001.

[12] Ecma. Standard ECMA-376, Office Open XML File Formats, December 2006.

[13] EPCglobal. EPCTM Tag Data Standards v. 1.1, Rev.1.24 Standard Specification 01, April 2004.

[14] EPCglobal. EPC Information Services (EPCIS) v. 1.0 Specification, April 2007.

[15] D. Fensel, H. Lausen, J. de Bruijn, M. Stollberg, D. Roman, and A. Polleres. *Enabling Semantic Web Services: The Web Service Modeling Ontology.* Springer-Verlag, Heidelberg, Germany, 2006.

[16] WAP FORUM. UAProf User Agent Profiling Specification, 1999, amended 2001.

[17] T.R. Gruber. A translation approach to portable ontology specifications. *Knowl. Acquis.*, 5(2): 199–220, 1993.

[18] T. Gu, X. Wang, H. Pung, and D. Zhang. An ontology-based context model in intelligent environments. In *Proceedings of Communication Networks and Distributed Systems Modeling and Simulation Conference*, San Diego, CA, January 2004.

[19] M. Handley, H. Schulzrinne, E. Schooler, and J. Rosenberg. SIP: Session Initiation Protocol, March 1999.

[20] I. Horrocks, P.P. Schneider, and F. van Harmelen. From SHIQ and RDF to OWL: The making of a web ontology language. *J. Web Semantics*, 1(1): 7–26, 2003.

[21] R. Hull. Converged Services: A Hidden Challenge for the Web Services Paradigm. In E. Bertino et al., eds. *EDBT*, vol. 2992 *of Lecture Notes in Computer Science*, pp. 1–2. Springer-Verlag, Heidelberg, Germany, 2004.

[22] International Telecommunication Union. *ITU Internet Reports 2005: The Internet of Things 2005*, 7th ed., Geneva, Switzerland, November 2005.

[23] International Telecommunication Union. *ITU Internet Reports 2006: digital.life*, Geneva, Switzerland, December 2006.

[24] G. Klyne, F. Reynolds, C. Woodrow, H. Ohto, J. Hjelm, M.H. Butler, and L. Tran. Composite Capability/Preference Profiles (CC/PP): Structure and Vocabularies 1.0, 2003.

[25] D. Martin, M. Burstein, J. Hobbs, O. Lassila, D. McDermott, S. McIlraith, S. Narayanan, M. Paolucci, B. Parsia, T. Payne, E. Sirin, N. Srinivasan, and K. Sycara. OWL-S: Semantic Markup for Web Services, Release 1.1, November 2004.

[26] OASIS. Open Document Format For Office Applications (OpenDocument) v. 1.0, May 2005.

[27] OASIS. Reference Model for Service-Oriented Architecture, August 2006.

[28] M. Philipose, K.P. Fishkin, M. Perkowitz, D.J. Patterson, D. Fox, H. Kautz, and D. Hahnel. Inferring activities from interactions with objects. *IEEE Perv. Comp.* 3(4): 50–57, 2004.

[29] D. Preuveneers and Y. Berbers. Adaptive context management using a component-based approach. In Nancy Alonistioti and Lea Kutvonen, eds., *Proceedings of 5th IFIP International Conference on Distributed Applications and Interoperable Systems (DAIS2005)*, Lecture Notes in Computer Science (LNCS). Springer-Verlag, Heidelberg, Germany, June 2005.

[30] D. Preuveneers, J. Van den Bergh, D. Wagelaar, A. Georges, P. Rigole, T. Clerckx, Y. Berbers, K. Coninx, V. Jonckers, and K. De Bosschere. Towards an extensible context ontology for Ambient Intelligence. In P. Markopoulos, et al. eds., *Second European Symposium on Ambient Intelligence*, vol. 3295 of *LNCS*, pp. 148–159, Eindhoven, The Netherlands, November 2004. Springer-Verlag, Heidelberg, Germany.

[31] N.S. Ryan, J. Pascoe, and D.R. Morse. Enhanced reality fieldwork: The context-aware archaeological assistant. In V. Gaffney, M. van Leusen, and S. Exxon, eds., *Computer Applications in Archaeology 1997*, British Archaeological Reports, Tempus Reparatum, Oxford, October 1998.

[32] B.N. Schilit and M.M. Theimer. Disseminating active map information to mobile hosts. *Network, IEEE*, 8(5): 22–32, 1994.

[33] B.N. Schilit, N.I. Adams, and R. Want. Context-aware computing applications. In *Proceedings of the Workshop on Mobile Computing Systems and Applications*, pp. 85–90, Santa Cruz, CA, December 1994. IEEE Computer Society, Washington, D.C.

[34] K. Sivashanmugam, K. Verma, A. Sheth, and J. Miller. Adding semantics to Web services standards. In *Proceedings of the 1st International Conference on Web Services (ICWS'03)*, pp. 395–401, Erfurt, Germany, June 2003.

[35] J.R. Smith, K.P. Fishkin, B. Jiang, A. Mamishev, M. Philipose, A.D. Rea, S. Roy, and K. Sundara-Rajan. RFID-based techniques for human-activity detection. *Commun. ACM*, 48(9): 39–44, 2005.

[36] V. Stanford. Pervasive computing goes the last 100 feet with RFID systems. *IEEE Perv. Comp.*, 2(2): 9–14, 2003.

[37] T. Strang, Claudia L. Popien, and Korbinian Frank. CoOL: A context ontology language to enable contextual interoperability. In J.B. Stefani, I. Dameure, and D. Hagimont, eds., *LNCS 2893: Proceedings of 4th IFIP WG 6.1 International Conference on Distributed Applications and Interoperable Systems (DAIS2003)*, vol. 2893 of *Lecture Notes in Computer Science (LNCS)*, pp. 236–247, Paris, November 2003. Springer-Verlag, Heidelberg, Germany.

[38] X.H. Wang, D.Q. Zhang, T. Gu, and H.K. Pung. Ontology-based context modeling and reasoning using OWL PERCOMW, p. 18, Second IEEE Annual Conference on Pervasive Computing and Communications Workshops, 2004.

[39] R. Want. Enabling ubiquitous sensing with RFID. *Computer*, 37(4): 84–86, 2004.

[40] S. Weerawarana, F. Curbera, F. Leymann, T. Storey, and D.F. Ferguson. *Web Services Platform Architecture: SOAP, WSDL, WS-Policy, WS-Addressing, WS-BPEL, WS-Reliable Messaging, and More*. Prentice Hall, New York, March 2005.

Index